Solar Radiation Data from Satellite Images

Solar Energy R&D in the European Community

Series F:

Solar Radiation Data

Volume 4

Publication arrangements: T. C. Jones

Solar Energy R&D
in the European Community

Series F Volume 4
Solar Radiation Data

Solar Radiation Data from Satellite Images

Determination of Solar Radiation at Ground Level from Images of the Earth Transmitted by Meteorological Satellites

by

W. Grüter, H. Guillard, W. Möser,
J. M. Monget, W. Palz, E. Raschke,
R. E. Reinhardt, P. Schwarzmann, and
L. Wald

D. Reidel Publishing Company
A MEMBER OF THE KLUWER ACADEMIC PUBLISHERS GROUP
Dordrecht / Boston / Lancaster / Tokyo

for the Commission of the European Communities

Library of Congress Cataloging in Publication Data

CIP

Main entry under title:

Solar radiation data from satellite images.

(Solar energy R & D in the European Community. Series F, Solar radiation data; v. 4)
Bibliography: p.
1. Solar radiation–Observations. 2. Solar radiation–Remote sensing. 3. Astronautics in meteorology. I. Grüter, W. II. Commission of the European Communities. III. Series.
QC911.8.S65 1986 551.5'271 86-478
ISBN 90-277-2204-8

Publication arrangements by
Commission of the European Communities
Directorate-General Information Market and Innovation, Luxembourg

EUR 10375

Published by D. Reidel Publishing Company
P.O. Box 17, 3300 AA Dordrecht, Holland

Sold and distributed in the U.S.A. and Canada
by Kluwer Academic Publishers,
190 Old Derby Street, Hingham, MA 02043, U.S.A.

In all other countries, sold and distributed
by Kluwer Academic Publishers Group,
P.O. Box 322, 3300 AH Dordrecht, Holland

Printed in The Netherlands

Contents

PREFACE

The present book was initiated within the Commission's Solar Energy R&D Programme. It provides an overview of the possibilities to determine solar radiation on the ground level from satellite images. Together with three important research laboratories, a new set of methodologies were produced and validated after many years of development work.

The methods which are described in this book will enable one to determine solar energies on the ground level with an accuracy better than 10%. For most of the work, images from the European Meteosat were used but images from the American satellites NOAA were also taken into account. The methods are relatively cheap because computer times are reasonably low.

In this way, a new convenient route becomes accessible to determine solar radiation data quickly and cheaply almost everywhere on the globe. In particular, in certain countries where a good measuring network for solar radiation up to the quality standards of the World Meteorological Organization is not currently available, it will now become possible to produce reliable data as are needed for any solar energy application. It will also become possible to set up special solar radiation data sets in the frame of particular R&D programmes as they are being implemented, for instance, in the pilot and demonstration programmes of the European Community.

It is interesting to note that the weather offices in Germany and France have already shown their interest in these new methods. They are considering their utilization in their general weather forecasting system.

This book comes as a valuable complement to the other important results of the Commission's R&D activities on solar radiation data; in particular, the two volumes of the European Solar Radiation Atlas which was published in 1984 and the Test Reference Years which have just become available in the form of data tapes for the purpose of computer calculations.

I take this opportunity to thank in particular Professor Raschke of the University of Cologne for finalizing the manuscript of this book. Important contributions were also made by Mr. Wald from Ecole des Mines in Sophia Antipolis. I would also like to thank all the other contributors and in particular the European weather services who provided the data for the validation exercise.

Dr. W. Palz
Head of the Solar Energy R&D Programme

PART I: SOLAR RADIATION

1.1 Solar radiation

1.1.1 General information and definitions

With the exception of geothermal energy, all forms of energy which are available within the Earth-atmosphere system are the result of the transformation of solar radiation. This radiation comprises wavelengths between 0.2 and 4 µm. Figure 1.1 shows the spectral distribution of the incoming solar radiation, which approximately corresponds to a black-body spectrum at 5800 K.

As shown in Fig. 1.1, about 40% of the solar energy reaching the top of the atmosphere is in the visible part of the spectrum (0.4-0.67µm). Before reaching the Earth's Surface, the radiation undergoes several processes within the atmosphere, namely: absorption and scattering by molecular particles and aerosols, and by clouds. These processes will be described in more detail in § 1.1.3. As a result, the downward solar radiation reaching the ground, called global irradiance (Dogniaux et al. 1984), is composed of two components:

-the diffuse radiation coming from the sky dome
-the direct radiation coming directly from the Sun.

If I_c is the direct irradiance at a plane perpendicular to the solar rays for clear sky and D the diffuse irradiance from the sky, the global irradiance G at a horizontal plane is:

$$G = D + I_c \cos \rho = I_c \sin \gamma + D \tag{1.1}$$

Where γ and ρ are the solar elevation angle and the solar zenith angle, respectively (see Fig. 1.2).

Parts of the radiation are reflected upwards and together with the radiation scattered back from the atmosphere, reach the satellite sensors. The purpose of the works described here is to infer the global solar irradiance G at a horizontal plane surface.

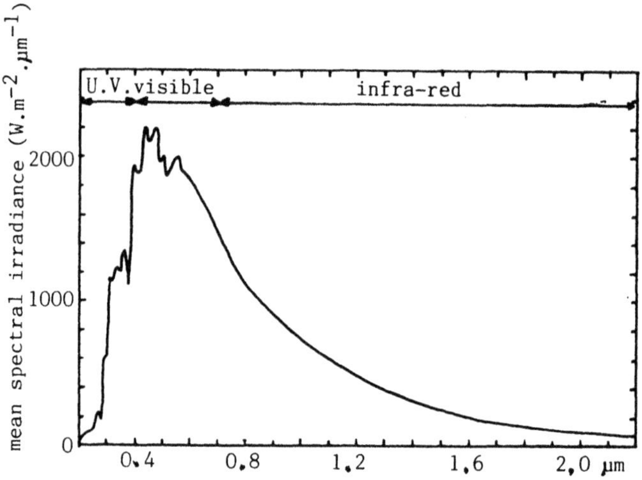

Fig. 1.1: Special distribution of the solar irradiance outside the atmosphere

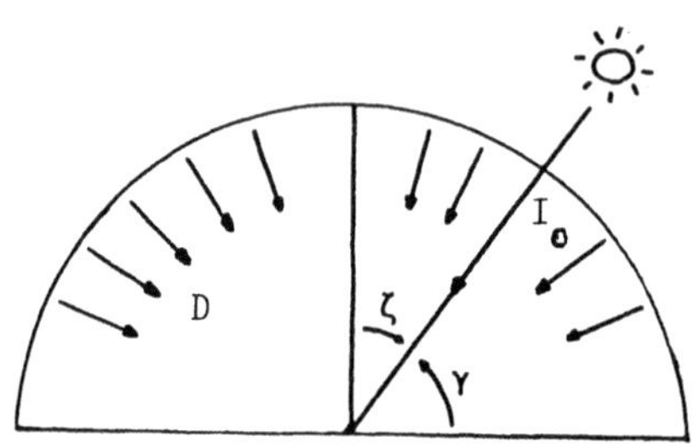

Fig. 1.2: Scheme of the different components of the global radiation at ground level. D = diffuse sky radiation; I_c = direct solar radiation

The terminology employed in the following chapters will be explained here. (The symbols are according to Dogniaux et al., 1984).

The radiance (units: $W\ m^{-2}\ sr^{-1}$) is defined as the radiant flux through a unit solid angle reaching an extended target in a given direction for unit projected target area in that direction (see Fig. 1.3). This is the quantity measured by a satellite radiometer with a narrow aperture.

The irradiance (units: $W\ m^{-2}$) is the radiant flux received at a given surface divided by the area of the plane (see Fig. 1.4). Note that the direction of the incident flux is not specified. The radiation flux emerging from a plane surface is called exitance. Then the relationship between irradiance I and radiance L is

$$I = \int_0^{2\pi}\int_0^{\frac{\pi}{2}} L(\theta,\phi) \cos\theta \sin\theta d\theta d\phi \qquad W\ m^{-2} \tag{1.2}$$

where the integration is performed on a hemisphere, where θ and ϕ are the zenith and azimuth angles, respectively.
It is often assumed that the reflectors (ground, sea or cloud surfaces) are isotropic (or Lambertian), that is the radiance does not vary with the angles of incidence or exitance, θ, ϕ. In this case the relationship between irradiance and radiance is simply

$$E = L \cdot \pi \quad W\ m^{-2} \tag{1.3}$$

We shall also make use of the concept of albedo (ρ) which is defined as the amount of upward radiance flux density divided by the downward irradiance (see Fig. 1.4).

Over a non-transparent surface this quantity is equal to the reflectance which is the ratio of reflected irradiance to incident irradiance. Therefore these terms are often employed indiscriminately, but at the sea surface or over optically thin clouds, these quantities are not identical, since there the upward radiation contains also components from below the sea surface or from below and within the clouds.

We shall conclude on terminology by defining the reflectance factor which is defined as the ratio of the radiance into the direction (θ', ϕ') and the radiance reflected in the same direction by a perfect reflecting diffuser.

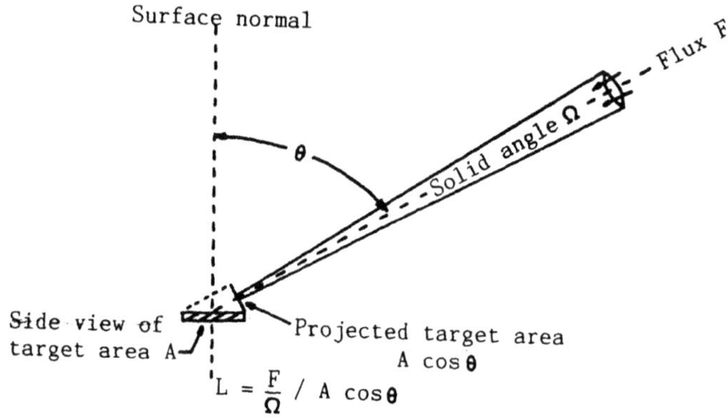

Fig. 1.3: Concept of the radiance L

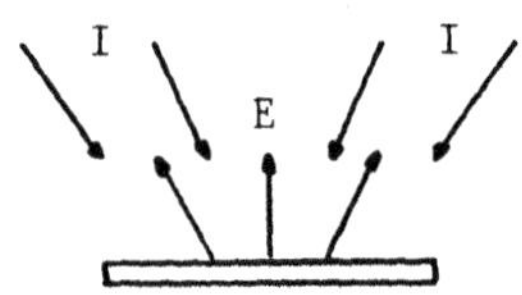

Fig. 1.4: Concept of the albedo

Since a perfect reflecting diffuser will totally reflect the total solar irradiance or global irradiance , G (θ_o, ϕ_o) the reflectance factor is equal to

$$R\ (\theta_o, \phi, \theta', \phi') = \frac{L(\theta, \phi').\pi}{G(\theta, \phi)} \qquad (1.4)$$

Satellite measurements are often expressed in this form. If the target is assumed to be Lambertian, R is a constant, whereas in reality R might show pronounced angular dependencies. For instance, over the ocean a strong specular component, the "sun-glint", can be observed.

1.1.2 Astronomical factors

The solar irradiance received outside the atmosphere at a plane perpendicular to the Sun's rays varies with the square of the Earth-Sun distance. Since the Earth's orbit is elliptical rather than circular, the actual energy received varies seasonally by $\pm$ 3.5 % around a mean value which is called the solar constant I_0 and is equal to 1370 W m ^{2} $\pm$ 0.1 % (Hickey et al. 1982).
However, to the universal tilt of the Earth's axis (23°27' to -23° 27') and the Earth's orbit around the Sun, the Sun's radiation at ground level undergoes strong diurnal and seasonal changes; also the geographical coordinates of each location must be taken into account. Respective values are listed in many handbooks (e.g.

1.1.3 Solar radiation on the ground

Knowing the time of the year and the latitude, it is straightforward to compute the solar irradiance reaching a horizontal surface at the top of the atmosphere. This is not the case for the solar irradiance at the Earth's surface. Before reaching the ground, the solar radiation undergoes absorption and scattering processes in the atmosphere, which considerably reduce the available energy at the ground (or sea) surface. A short description of these processes follows:

- molecular (Rayleigh) scattering which approximately follows a λ^{-4} law where λ is the wavelength.

- selective absorption by atmospheric gases within definite wavelength ranges, in particular O_3, O_2, CO_2, H_2O. Fig. 1.5 illustrates the effects

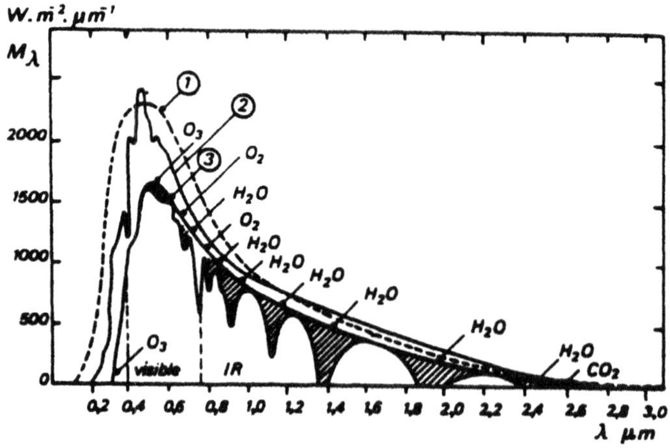

Fig. 1.5: Spectral distribution of the solar radiation at ground level (cloudless sky);
1: Black body irradiance at 5800 K;
2: Spectrum outside the atmosphere;
3. Spectrum at sea level. Hatched areas are absorption bands.

of these absorptions on the solar spectrum.

- scattering and absorption by aerosols in the atmosphere
- reflection and absorption by clouds has a dominant effect on the radiation fields.

Fig. 1.6 shows the average contribution of these different processes.

The methods which are used to estimate the global radiation at ground level must take careful account of the optical transfer properties of clouds. Some of them will be discussed in the following chapter.

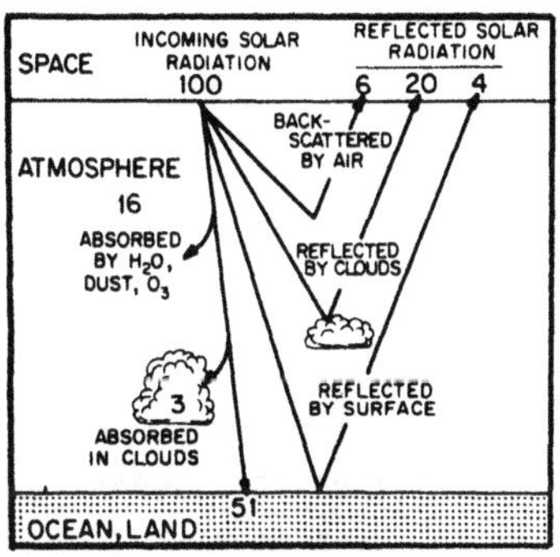

Fig. 1.6: Schematic diagram of radiation processes in the atmosphere. The incoming solar radiation at the top of the atmosphere is given in 100 relative units.

1.2 Measurements of solar radiation

1.2.1 Classical methods

At a few stations, the following parameters are measured at ground level: sunshine duration and the various components of the solar radiation (global, direct, diffuse and possibly also spectral). At most stations, only the global radiation or the sunshine duration are measured. An excellent description of basic instruments has been summarized by Coulson (1975).

The sunshine duration S is measured using a heliograph. It records the time span during which the direct solar irradiance is greater than a certain threshold of about 100 to 150 Wm^{-2}. This type of measurement is performed each day at many meteorological stations. This quantity is often reduced to the maximum sunshine duration S_{max} thus leading to the sunshine fraction ratio $\gamma = S/S_{max}$ (Dogniaux et al.). Several different instrumental designs are in use.

The direct solar radiation E can be measured with a pyrheliometer. The diffuse component D of the solar radiation at ground level can be measured with a pyranometer where a canopy or ring shields the direct components. This method requires correction factors which depend on the time of day, cloud cover, etc. Research on these correction factors is in progress (e.g. Dehne, 1980). The total solar radiation flux to ground, or the global irradiance, is now usually measured with a pyranometer. Various designs are in use, although it is known that their measurements might deviate by up to 3-5 per cent (cf. Ebke, 1982). Fig 1.7 shows the pyranometer network in several European countries (status in 1980), which are members of the EEC. This network has been increased by many stations during the last years.
Finally, let us note that cloud cover is usually reported by meteorological observers on a numerical scale ranging from 0 to 10 or from 0 to 8. In many areas especially over the oceans, this quantity is often the only one available on the transparency of cloud fields for solar radiation.

1.2.2 Empirical methods

Taking into account the difficulty of obtaining reliable measurements of solar irradiance, numerous authors use all available quantities, primarily the cloud cover and the sunshine duration, to estimate the global irradiance at the Earth's surface.

As clouds are the most efficient reducers of the solar radiation, numerous works have tried to correlate solar irradiance at the Earth's surface with the cloud cover.

Most methods are based on the preliminary determination of clear sky irradiance. Among the methods proposed, a regressional relation of measurements with some parameters on the Sun's position is one of the simplest. Observations are fitted with the following relation:

$$G_c = a_s (\sin \gamma)^{b_s} \tag{1.5}$$

where γ is the solar elevation angle and a_s and b_s are coefficients related to the mean conditions of sky brightness and colour. G_c = clear sky global irradiance.

The solar radiation under cloudy conditions is then estimated by introducing a correction function related to the cloudiness. Table 1 summarizes some of the numerous formulae proposed:

Table 1 Some empirical formulae of various authors to estimate the global radiation. C is cover of low clouds, Φ is geographical latitude. G_c = clear sky global irradiance.

KIMBALL	1928	$G=G_c(1-0.71C)$
MOSBY	1936	$G=G_c(1-0.71C)$
GABITES	1950	$G=G_c(1.04-0.71C)$
HOUGHTON	1954	$G=G_c(1.28-1.03C)$

These empirical formulae are valid only for monthly or annual means. For hourly or daily estimates, other formulae have been proposed (Table 1.2).

Table 1.2 Empirical formula to estimate daily means of the global irradiance.

C is cloud cover in 1/10, C_l is fractional cover of low clouds, C_m of medium clouds, and C_h of high cloud;
$G_c = 2 E_o \cdot /1-(0.371 \sin\gamma + 0.225 \sin 3\gamma + 0.0842 \sin 5\gamma + ...)/$ where E_o is the solar constant, and is the Sun's elevation; G_c = clear sky global irradiance.

BUDYKO	1956	$G=0.94\, G_c(1-0.68C)$	all clouds
BERLIAND	1960	$G=G_c(1-0.65\, C^2)$	all clouds
LAEVASTU	1965	$G=G_c(1-0.0006\, C^3)$	all clouds
		$G=G_c(1-0.0075\, C_l^{\,2})$	for low clouds
		$G=G_c(1-0.006\, C_m^{\,2})$	for medium clouds
		$G=G_c(1-0.0035\, C_h^{\,2})$	for high clouds
MASUIKE, MORINAGA	1970	$G=G_c(1-0.026\, C^{1.3})$	all clouds

However, estimates of the cloud cover in terms of tenths or eighths can be very subjective and vary with the observer. The sunshine duration is an almost objective quantity and is measured at meteorological stations. Therefore, numerous investigations, mainly carried out by agrometeorologists, have been attempted to obtain empirical relations between the global radiation and the sunshine duration. Other relationships relate the sunshine duration to the ratio between the observed global irradiance and the clear sky global irradiance, as originally proposed by Angström in 1924. There have also been attempts to determine the total transmission factor (transmittance) which is defined as the ratio between the global irradiance on the ground and the extraterrestrical solar irradiance. Some these empirical formulae are summarized in Table 1.3.

Table 1.3 Empirical formulae for estimating global irradiance from the sunshine duration. G_o is the irradiance of a horizontal plane outside the atmosphere. G_c is clear sky irradiance. S_h is the hourly sunshine duration, and a, b, z are empirical constants.

ANGSTRÖM	1924	$G=G_c\ (0.25+0.75S/S_{max})$
KIMBALL	1927	$G=G_c\ (1-0.78\ S/S_{max})$
MCDONALD, FRITZ	1949	$G=G_c\ (0.35+0.61\ S/S_{max})$
BOURGES	1979	$G=G_c\ (a + b\ S_h + z)$

The first three formulae are for monthly or decade means. This kind of formula is currently used by agronomists with variable constants for the computation of the heat balance at ground level. The last formulae (Bourges, 1979) has been derived to estimate hourly averages and is used for solar energy conversion systems where a statistical distribution of hourly values of the global radiation is needed.

Despite the wide dispersion that one can note in the formulae listed in Tables 1.1, 1.2 and 1.3, on average they provide estimates of the solar irradiance with a relative accuracy of about 20 % or better. However, it must be emphasized that the empirical regression coefficients depend greatly on the locations and seasons.

1.2.3 The potential usefulness of remote sensing data

The density of the radiation measurement network at ground level in Europe is still low and the average distance between stations is around 150-200 km. This is not sufficiently refined to take into account smaller scale variations which are important for the use of solar energy.
In other parts of the world, the networks are even more sparse.

On the other hand, meteorological satellites provide observations of cloud fields over the whole surface of the Earth, usually in two spectral ranges: the visible and thermal infrared, with a ground resolution of 1 km to 5 km depending on the instrument and the latitude of measurement. The time interval of these observations varies between 12 h and 30 minutes.

Proper processing of these new imaging data would certainly provide a wealth of information useful in the production of solar atlases, particularly for those areas where no traditional observations are available. But it can also be expected that these satellite data will make it possible to investigate spatially small-scale variations of available solar energy. Knowledge of which is needed for planning purposes. Therefore the Solar Energy Research and Development Programme of the European Community has carried through a development project (Project F, Action 4.3, 1980-1983) aimed at the determination of radiation measurements where satellite data provide the major input.

It should be emphasized here that these data on the spatial and temporal variability of the global radiation at ground level are also of great scientific value. Many investigations require accurate data on the heat transfer into the oceans, which is primarily provided by the Sun.

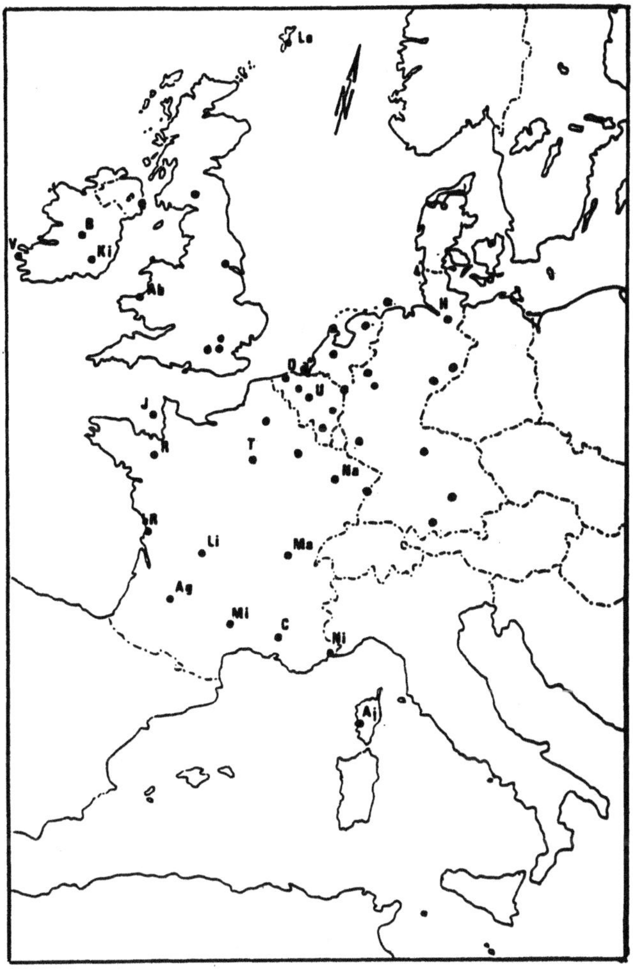

Fig. 1.7: Location of pyranometer stations used by the project in Europe (status at about 1980).

PART II: METEOROLOGICAL SATELLITES

2.1 Description

During the past decade, several meteorological satellite systems have been launched, of which some are still in operation. Of particular relevance for our purposes is the geostationary system (METEOSAT, GOES, GMS, INSAT) and the polar orbiting system of the NOAA series. The Earth-observing payloads of these systems are different in many features and are described below in more detail.

2.1.1 The geostationary satellite system

Within the framework of the World Weather Watch five geostationary satellites at least are required to monitor permanently the major part of the Earth. The present system consists of the METEOSAT, INSAT, GMS, and GOES satellites.

These satellites are orbiting at an altitude of about 36,000 km above sea level with the same angular speed and direction as the Earth, so that they remain at a fixed position relative to the Earth over a given location at the equator. The European METEOSAT satellites are part of this observational network.

METEOSAT 1 was launched on 23 November 1977 and its imaging radiometer failed operations only two years later. It was then replaced by METEOSAT 2, launched on 19 June 1981. The location of the sub-satellite point of both satellites was (0°N ; 0°E); but METEOSAT 1 is now drifting away (Status: Jan. 1985).

The imaging radiometer aboard the METEOSAT satellite has three channels:

- infrared channel ("water vapour")	5.7 - 7.1 μm
- infrared channel (surface temperature):	10.5 - 12.5 μm
- solar channel	0.4 - 1.1 μm

The spatial resolution near the subsatellite point is 2.5 km for the solar channel and 5 km for the other two. The pixel size increases when the zenith angle increases. Over Europe the solar pixel sizes cover areas of 4-6 km in diameter. The data are first received at the ground station in Darmstadt. Then the raw data are pre-processed and again relayed by the METEOSAT satellite for recording by primary and secondary user stations within the satellite's field of view. In the latter case the transmission format is called WEFAX.

2.1.2 The NOAA satellites

The operational NOAA satellite system consists of two satellites in a sun-synchronous near polar orbit, and in opposite phase.

The orbit characteristics as well as the payload are slightly different from one satellite to another. The inclination of the orbits is about 99° and the period is about 102 minutes per orbit. The mean altitude is 850 km.

The payload of the NOAA satellites consists of several systems of instruments:

- TIROS Operational Vertical Sounder (TOVS) comprising
  - High Resolution Infrared Radiation Sounder
  - Stratospheric Sounding Unit
  - Microwave Sounding Unit

- Data Collection System (DCS)

- Space Environment Monitor (SEM) comprising
  - Total Energy Detector
  - Medium Energy Proton - Alpha Detector
  - High Energy Proton - Alpha Detector

- AVHRR: Advanced Very High Resolution Radiometer

For solar meteorology purposes only this last instrument appears to be relevant. The AVHRR provides spectrally low, but spatially very highly resolved images. It is scanning the Earth surface continuously, perpendicular to the subsatellite track.

Owing to the scanning geometry, the image data are distorted twofold due to: non-linear change of pixel size with increasing nadir angle (curvature effect), and the daily revolution of the Earth.

The scan of the radiometer is performed simultaneously in the different spectral channels. During the backscan of the instrument, calibration data are inserted into the data flow.

The spectral response of the AVHRR varies only slightly from one satellite to another. The spectral windows of the AVHRR aboard NOAA 7 and 6 are listed below. Meanwhile (Jan. 1985) already NOAA 9 is in operation with an almost identical AVHRR.

		NOAA-7	NOAA-6	Remarks
Channel	1	0.58 - 0.68 μm	0.58 - 0.68 μm	Reflected solar
	2	0.725 - 1.1 μm	0.725 - 1.1 μm	radiation
	3	3.55 - 3.93 μm	3.55 - 3.93 μm	Reflected solar and emitted terrestrial radiation
	4	10.3 - 11.3 μm	10.5 - 11.5 μm	To measure surface temperatures
	5	11.5 - 12.5 μm	-	
		(channel 5 not on all satellites)		

Channel 1 covers parts of the visible region. Channels 2 and 3 are in the near infrared range whereas channels 4 and 5 lie in the far infrared region. All spectral ranges have been chosen within "atmospheric windows" to minimize the interference by emission and absorption due to atmospheric gases.
Three groups participated in action 4.3 of Project F of the Solar Energy Programme of the Commission of the European Communities. These are the Universities of Cologne, Sophia - Antipolis and Stuttgart. Their data sources are summarized in Table 2.1.

Table 2.1 Types of satellite data processed in Action 4.3 by the research groups.

Group:	COLOGNE	SOPHIA-ANTIPOLIS	STUTTGART
Type of satellite	METEOSAT geostationary	METEOSAT geostationary	NOAA polar orbiter
Time sampling	30 min	30 min	6 - 12 h
Resolution	2.5 km Solar 5 km IR	2.5 km Solar	1 km
Spectral Properties	Solar: 0.4-1.1 µm IR: 10.5-12.5µm	Solar: 0.4-1.1 µm	CH1: .55 - .68 µm CH2: .725- 1.1 µm CH3: 3.55 - 3.93 µm CH4: 10.3 -11.3 µm CH5: 11.5 -12.5 µm
Availa-bility	Operational programme (up to 1990) World wide coverage		Operational (up to 1990) World wide coverage
Remark	Available through Darmstadt, or PDUS and SDUS broadcasting		- Spectral bands subject to change -Two satellites operational at one time.

2.2 Interpretation of satellite imagery

2.2.1 Emission and reflection processes

If subjected to a radiation stream a body will transmit a fraction τ, absorb a fraction α and reflect a fraction ρ. If this body is in thermal equilibrium, the absorbed portion α will be radiated away by an emission process. The following relation holds:

$$\alpha(\lambda) + \rho(\lambda) + \tau(\lambda) = 1$$
$$\alpha(\lambda) = \varepsilon(\lambda) \qquad (2.1)$$

where λ is wavelength
α is spectral absorptance
ε is spectral emittance
ρ is spectral reflectance
τ is spectral transmittance

A perfect black body is a body for which the spectral emittance is unity, while a perfect reflector has a reflectance of unity.

The intensity or radiance of the emission of a body is a function of the wavelength and of its temperature T and is given by:

$$L(\lambda, T) = \varepsilon(\lambda)\, B(\lambda, T) \qquad (2.2)$$

where B (λ, T) is the intensity of the black body at temperature T and is determined by Planck's law:

$$B(\lambda, T) = \frac{2\,hc^2}{\lambda^5 \exp\left(\frac{hc}{kt} - 1\right)} \qquad (2.3)$$

where h is Planck's constant
k is Boltzmann's constant
c is speed of light in a vacuum

The wavelength for which the spectral intensity $B(\lambda, T)$ reaches its maximum is a function of T and increases when the temperature decreases (Fig. 2.1) (Wien's law). The wavelength of maximum intensity is located at about 10µm for a temperature of + 27° C and at about 13µm for a temperature of about - 33° C.

2.2.2 Images in the infrared and solar range

In images taken in the solar spectrum one observes the radiation reflected by the ground or by the clouds. Thus if it is assumed that the reflection is isotropic or Lambertian, the solar image is strongly related to the albedo of the ground and clouds. The albedo of thick clouds is very high, ranging from 0.6 to 0.9. It is of the same magnitude as the albedo of snow and ice, but it is higher than the albedo of common continental surfaces (less than 0.3) and much higher than that of the ocean (about 0.05). Note of that the ocean reflectance is not at all Lambertian and since the radiometer measures radiances and not irradiances, the bidirectional reflectance or reflectance factor concept is more relevant to the ocean rather than the albedo concept. Thin or broken cloud fields appear often only slightly brighter than the clear background and are hardly observable.

To summarize: In the solar images, the clouds will appear very bright together with the snow and parts of deserts while all water surfaces (ocean, lakes and rivers) will be the darkest parts in the images, whereas bare and vegetated soils will appear in intermediate tones.

In the infrared image, which nearly reveals the temperature of the observed bodies, usually the temperature of the cloud surfaces is very low so that they can be discriminated from the warmer ground signal. However, thin clouds enable a transfer of a non-negligible part of the radiation emitted from the ground below. Then the observed temperature contains information from both sources; its values range between the temperature of the top of the cloud and the temperature of the ground below. Such clouds (e.g. cirrus) will then appear much warmer and might be subject to misinterpretation. Further clouds on top of inversion layers are even warmer than clear areas nearby. This phenomenon can be observed very often over both polar regions.

These statements are illustrated in Fig. 2.2, which shows a bidimensional histogram of simultaneous measurements in the solar and infrared spectral ranges.

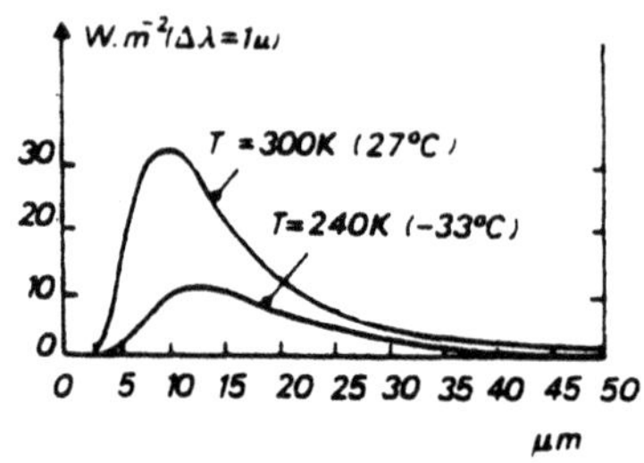

Fig. 2.1: Spectral intensities emitted by the black body at temperatures 300 K and 240 K

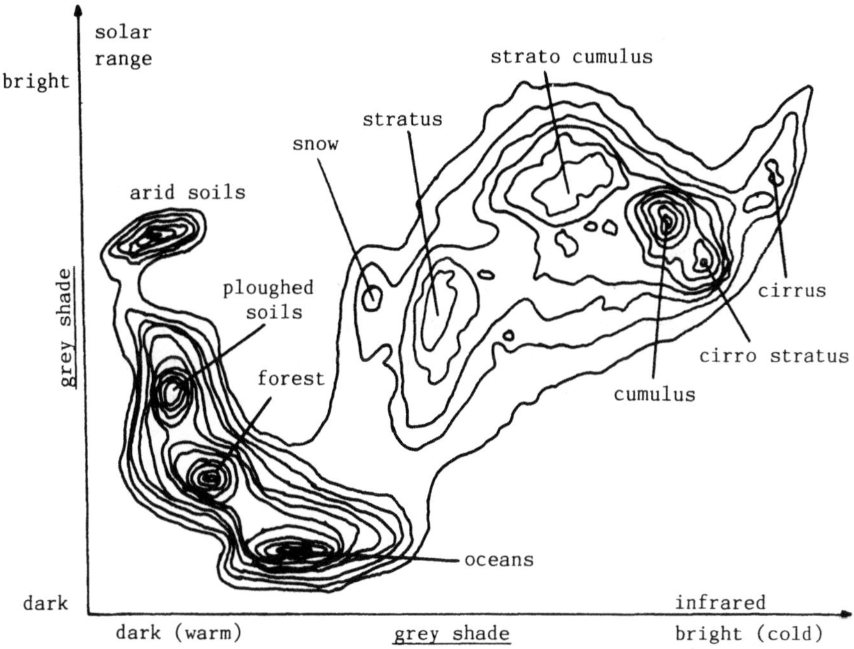

Fig. 2.2: Bidimensional histogram of the solar (y axis) and infrared (x axis) radiances, from Gube (ESA-Journal, 4, 381-396, 1980)

2.3 Bibliographical study on estimates of solar radiation from satellite data

The methods which use satellite data to obtain estimates of the solar irradiance at ground level can be split formally into two categories. In the first category, use is made of statistical relationships, between satellite-borne observations and solar ground measurements, while in the second category, physical radiative transfer models are used to derive a relationship between satellite measurements and solar irradiance at ground level.

While perhaps more precise when applied to small areas, statistical methods are less general and need careful adjustment for each location. On the other hand, since satellites can at present measure only a few parameters among the many that affect the radiative transfer through the atmosphere, the "physical" methods require many model assumptions. They are useful for all situations. Some averaging over space and time permits better results to be obtained but at the cost of degrading the spatial and temporal resolution. Finally let us note that the methods have to take into account the calibration of the satellite radiometer which is by no means a negligible problem.

2.3.1 Statistical Methods

Among the methods which belong to this category, the best known are those of Tarpley, Pastre and Hay (1980)

Tarpley (1979) used data from the geostationary satellite GOES 1 to obtain hourly and daily estimates of the solar irradiance in the Great Plains during summer of 77. The test region has been divided into subtargets of 50 km on each side, corresponding to averages of 7x6 picture elements (pixels). The method is based on the definition of a fractional cloud amount which is correlated with measurements from the pyranometer network. In a preparatory step, the target brightness B under clear sky conditions was determined by fitting the following equation:

$$B = a+b \cos\theta_o + \sin\theta_o \cos\phi_o + d \cdot \sin\theta_o \cos^2 \phi . \qquad (2.4)$$

where a, b, c, d are regression coefficients and θ_o and ϕ_o are the local solar zenith angle and the azimuth angle between sun and satellite. This has been done with a time series of satellite data collected during May 1977 just prior to the test period where cloudy data were eliminated by an iterative procedure. The clear sky brightness value predicted by equation (2.4) is then used to define two thresholds which can be attributed to three classes of cloud coverage: clear, 50% covered, or 100% cloud covered, and the fractional cloud amount N is defined on each subtarget by a weighted average of the number of pixels in each class. Depending upon the value of the cloud amount,three statistical relationships were derived between the fractional cloud amount and ground pyranometer measurements.

The results show a correlation coefficient of 0.91 for daily means with a relative dispersion of less than 10% when data from seven images per day were used. For hourly means, the correlation coefficient amounted to 0.94 with a relative dispersion of less than 10%. Figure 2.3 shows the results obtained for daily means.

Pastre (1981) estimated the sunshine fraction from the visible data of METEOSAT. This quantity can be converted into the solar irradiance at ground by empirical formulae of the form quoted in Section 1.2.2 which have been tested for the region of interest. In this work the cloud coverage is defined on each field element (the author chose averages of 3 x 3 pixels, corresponding to 20 x 20 km as the number of pixels whose values are above a definite threshold albedo) . A rough calibration of the satellite signal was done; the accuracy was estimated to be within 5 to 10%.

In principle this method is valid for long time periods, e.g. covering a month, and the repeat cycle of 30 min for METEOSAT is then judged to be accurate enough for this purpose. Figure 2.4 shows an example of the results obtained with this method.

2.3.2 Physical methods

In these methods the transmission of radiation through the atmosphere is simulated with algorithms which are based entirely on physical arguments. Their results are then used to derive a relationship between the global irradiation at the earth's surface and the quantities measured by the satellite (ie. the radiance in the solar and infrared spectral ranges).

There is now a large variety of numerical methods available to calculate the transfer of radiation in realistic models of the atmosphere inclusive of cloud layers. They are based on expansion of the radiation field in spherical harmonic functions (Dave and Canosa, 1974), or the use of matrix operators (Howell and Jacobowitz, 1970), of discrete ordinate methods (Liou, 1976), of iterative methods (Raschke, 1972) and of Monte Carlo techniques (Plass et al., 1968).

However, since the contribution of the atmosphere gases to the measurements in the above-quoted spectral ranges, which are all located in atmospheric windows, is very small very simplified models of the atmosphere can be used (Gautier et al., 1980; Dedieu et al., 1983; Raschke et al., 1982) which already yield satisfactory results.

As an example of these methods we describe the work of Dedieu et al (1983) which uses a very simple model of the atmosphere to analyse the data of METEOSAT.

A cloud layer of albedo $_c$ is assumed located above a surface of albedo ρ_s. The downward solar irradiance at the earth's surface I^-_s and the upward irradiance at the top of the atmosphere E^+_0 are given by:

$$I^-_s = I^-_0 \cdot (1 - A_c) / (1 - A_c \cdot A_s) \tag{2.5}$$

$$E^+_0 = A\, E^-_0 \tag{2.6}$$

where I^-_0 is downward solar irradiance at the top of the atmosphere ($I^-_0 = \cos\theta_0 \cdot E$, where E is the solar constant and θ_0 the Sun's zenith angle) and is the effective albedo of the earths atmosphere as measured by the satellite. It can be estimated from:

$$A = \rho_c n + \frac{\rho_s}{1} - \frac{(1-\rho_c' n)^2}{\rho_c \rho_s n} \qquad \frac{\rho_s (1-\rho_c n)^2}{1-\rho_c - \rho_s \cdot n} \tag{2.7}$$

Equations (2.5), (2.6) and (2.7) are combined to give a relation between the solar irradiance at the surface and the satellite measurements:

$$I_s = I_0 \frac{1-\rho}{1-\rho_s} \qquad (2.8)$$

The factor $\frac{1-\rho}{1-\rho_s}$ can be interpreted as transmittance of the atmosphere for downward diffuse and direct solar radiation measured at cloud free sky. The surface albedo ρ_s is determined. It should be noted that in a snow-covered area (where ρ_s is high) this relation must be used with caution.

Absorption and scattering by atmospheric gases can now be taken into account by introducing a clear sky transmission factor $\tau_c(\theta_o)$ in equation (2.8):

$$E_s = E_o^- \tau_c(0_o) \frac{1 - \rho/t(\theta_o)\ t\ (\theta)}{1 - \rho_s/\ t\ (\theta_o)\ t\ (\theta)} \qquad (2.9)$$

where 0 is the zenith angle of observation. Values for have been computed with relations derived by Lacis and Hansen (1974).

Equation (2.9) provides a linear relation between the solar irradiance at ground and the apparent albedo of the earth-atmosphere system. Assuming no angular dependence of ρ , then we can convert ρ into radiances L, as measured at satellite altitude by

$$A = \frac{L \cdot \pi}{E_o \cos \theta_o} \qquad (2.10)$$

In the work of Dedieu et al (1983) an additional problem arose from the use of METEOSAT data. Since solar channels of the radiometer of this satellite are uncalibrated, an accurate relation between the digital counts x and the radiance is not available but can be estimated from measurements over specific targets. Dedieu et al (1983) used a linear relation,

$$L = \underline{a}\ x \qquad (2.11)$$

where a value for $\underline{a}$ of 5.69 W m^{-2} sr^{-1} $count^{-1}$ is obtained as the best fit between the results of the model and the pyranometer data. This is not a very satisfactory procedure and a better way of calibration must be found. Fig. shows the results of a comparison with pyranometer data. The correlation coefficient is 0.85.

Although they use other physical models of the transmission of radiation through the atmosphere, the principles of the methods used by Gautier et al. (1980); Raschke and Preuss (1979) and Amado et al. (1981), Möser and Raschke (1984), are essentially similar to the ones described above. They are described in detail in the following PART THREE.

Gautier et al (1980) used two models: one for clear sky conditions and another for cloudy conditions. A threshold level is used to separate the cloudy pixels from the others. The calculation of the incoming irradiance is then performed with the appropriate model. Data from the GOES satellite were utilized after averaging the satellite data over an area of 8 x 8 pixels, which corresponds to an area of 10 x 10 km at the satellite subpoint.

We can also note from these studies that a better precision is attained for daily insolation than for hourly insolation. This is due to the essentially random nature of the cloud cover, the different time and spatial sampling of the satellite and of ground measurements, and also to the fact that the physical models describe the atmosphere in a statistical sense. They use already spatially degraded imagery, which makes the intercomparison with measurements from stations even more complicated.

The work of Amado et al. (1981) is devoted to the Sahel zone of Africa. Meteosat data are used with their maximal resolution of the satellite (2.5 x 2.5 km) for an area ranging from the Gulf of Guinea to the Sudan. Later the results are averaged over subareas of 4 x 4 pixels.

From a time series of only three days for the same slot (i.e. the same local hour), the minimum of the reflected radiation in cloudless situations as measured by the satellite was determined. In this case cloud-free situations have been met over all areas. This value is used as a threshold to determine cloud occurrence. In clear sky conditions, the incoming solar irradiance is computed with a model retaining absorption and diffusion coefficients dependent on the amount of water vapour in the atmosphere. In cloudy situations, three thresholds, determining 4 cloud types are used. For each type a separate absorption function has been utilized in allowing for the computation of solar irradiance at ground level.

Measurements from only 12 ground stations have been available to control the results. The comparison with them is not representative. For hourly insolation, the correlation coefficient reached values around 0.8.

A slightly different approach has been used by Raschke and Preuss (1979), with data of the NIMBUS 3 satellite. The results are obtained on 250 x 250 km areas. First a factor A is introduced:

$$A = \frac{A_p - A_{min}}{A_{max} - A_{min}} \tag{2.12}$$

where A_p is the mean monthly planetary albedo, and A_{min} is the minimum albedo (clear sky planetary albedo) determined as the minimum of a time series over some area; A_{max} is the maximum albedo computed from a theoretical cloud model. The downward flux at ground is then computed

$$F\downarrow = (1 - A)\, F_c\downarrow + A\, F_o\downarrow \tag{2.13}$$

where $F_c\downarrow$ and $F_o\downarrow$ are the flux densities of downward solar radiation at clear sky and in cloudy conditions. These quantities are calculated with numerical models for the radiative transfer in the atmosphere. A comparison with ground-based pyranometer measurements shows a non-negligible dispersion, which can be explained by the fact that these authors had satellite data only, which were already averaged over areas of 250 x 250 km. This method has been further explored by Möser and Raschke (1983, 1984). Tuzet et al. (1984) applied it also to measurements made over the Sahel zone.

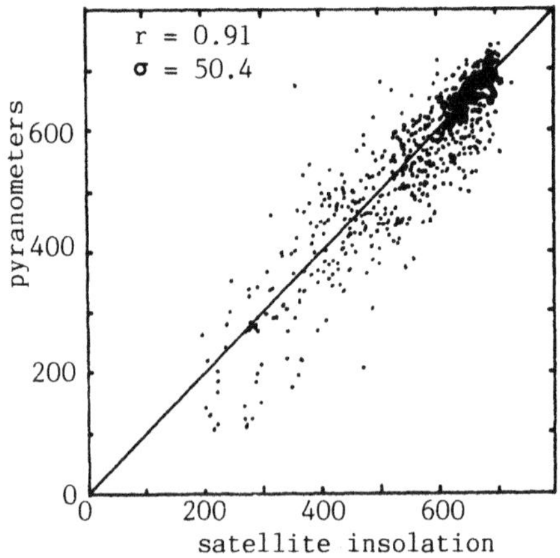

Fig. 2.3: Daily insolation measured by the 22 pyranometers (dependent data) plotted against satellite estimates (7 pictures per day).
Units are langleys. Diagonal line shows perfect agreement between pyranometer and satellite estimates (from Tarpley, 1979)

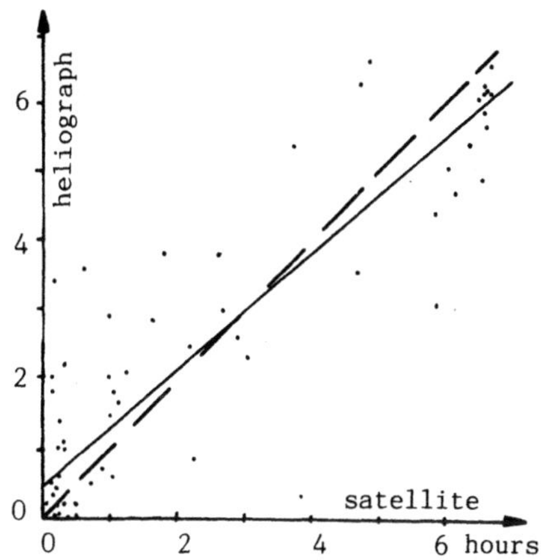

Fig. 2.4: Comparison between the sunshine duration as measured by a heliograph and estimated from satellite data (15 August 1979, afternoon from Pastre, 1981).

PART III: DETERMINATION OF SOLAR RADIATION FROM SATELLITE IMAGES (Action 4.3 of Project F)

3.1 Introduction

Under Project F of the European Community Solar Energy Research and Development Programme, Action 4.3 was dedicated to the development and intercomparison of algorithms which are used to determine global irradiance at ground level using satellite imagery. Within this programme, which extended from 1980 to 1983, three research groups have developed original methodological approaches to the processing of satellite data, based either on statistical data analysis or realistic physical models of the measurements.

The research institutions which were involved are located in Germany and France:

- Institut für Geophysik und Meteorologie, University of Cologne (E. RASCHKE, W. MÖSER)
- Institut für Physikalische Electronik, University of Stuttgart (E. REINHARDT, P. SCHWARZMANN)
- Centre de Télédétection et d'Analyse des Milieux Naturels (CTAMN), Ecole des Mines de Paris, Sophia-Antipolis, (J.M. MONGET, D. CANO).

As mentioned earlier, there are too few ground stations even in Europe to obtain a picture of the spatial and temporal variation of the global radiation at ground level which is clear and detailed enough to provide information for solar energy implementation projects. Therefore attempts have been sponsored to deduce the relevant information from satellite data. This promising new source of information has been assessed during the four-year period of Action 4.3. Each of the three participating research groups demonstrated the feasibility of their own solution to the problem. The following chapters describe in more detail the methodologies developed by these three research groups.

3.2 Method of Möser and Raschke (Cologne)

This method is based on the determination of the solar irradiance on ground as a function of the radiance measured by the satellite. So it may be classified as a physical approach according to § 2.3.2. The relations between the solar irradiance at the Earth's surface and the reflected radiation at the top of the atmosphere are derived from results of a radiative transfer model developed for this study. This model is more elaborate than the simple ones used by Gautier et al. (1980) or Dedieu et al. (1983). It permits the inclusion of more parameters in the description of the atmosphere. However, as only one or two parameters can be estimated by satellite measurements, the other parameters used in the model must be specified climatological means.

An additional problem arises from the fact that the data of the solar channel of METEOSAT used for this study are not calibrated, thus the absolute value of the measured radiance is unknown. Therefore, the group in Cologne had to choose an indirect procedure to incorporate these measurements of Meteosat into the model computations. Similar problems have been faced by the other two groups.

3.2.1 The model

A radiation transfer model which is based on a two-stream approximation was used. A complete description of this radiative transfer model can be found in Kerschgens et al. (1978). This method does not permit calculation of the azimuth and zenith dependencies of the radiation field. But it does permit study of the effects of various atmospheric parameters on the radiation field with little computational time.

In the following we summarize the main variables used in this model in the computation of the radiation field:

a) θ_0: zenith-distance of the sun

The angular range of interpretation should be limited to about 80° because of the plan-parallel layers in the model;

b) h_c: cloud top height

Each layer of the model can be a cloud layer, therefore the vertical extension and thickness of clouds is limited only by the vertical resolution of the model layers, which have been chosen as

about 200 m near ground, and

about 500 m in the upper troposphere;

c) δ: optical depth of the cloud

The range of definition of δ should be limited to 50-100. No further information is obtained when a larger optical thickness is used, since then model clouds do not change their radiative transfer properties.

d) Cloud type (droplet size distribution)

Droplet - size distributions are described here for stratocumulus (r_{eff} = 5.33 µm; V_{eff} = 0.118) and nimbostratus (r_{eff} = 10.81 µm, v_{eff} = 0.143) where r_{eff} is the mean effective droplet size and v_{eff} is the effective variance of the droplet-size distribution.

Although there is a strong difference in the transmittance of the two cloud types, the reflectance differs only a little. Therefore these authors assume in all further calculations an alto-stratus layer with the following effective values (r_{eff} = 7.01 µm, v_{eff} = 0.113).

e) Boundary layer structure

The atmospheric boundary layer may extend in their model up to a height of 2.0 km above ground. It is resolved in steps of 200 m, where the dust layer (i.e. layer with constant visibility) may extend up to a height of 1.6 km. Three types of aerosols are used to characterize maritime, rural and urban conditions.

f) Condition of the atmosphere above

These are climatological profiles of temperature, pressure, humidity and ozone. They are given with a resolution of 1 km.

g) Ground Albedo

A detailed knowledge of the ground albedo is required. Typical values are selected later on.

Calculations were performed for a large variety of atmospheric conditions. The results made it possible to derive relations between the global irradiance at the Earth's surface and the reflected radiation field depending on the atmospheric conditions. Results show, in concordance with observations, that cloudiness is the main factor affecting the radiation field and that other parameters like aerosol distribution, temperature, ozone, CO_2 or moisture profiles cause much smaller effects.In cloudy situations, the effects of droplet-size distributions, cloud top heights and of the optical thickness of the clouds are investigated. For practical purposes the effect of droplet-size distribution was judged negligible and the effects of the cloud top heights are of second order importance. In order to facilitate utilisation of the results on a minicomputer these authors approximated and stored them in the form of Tschebyscheff polynomials.

3.2.2 Principles of the method

Calculations performed with the two-stream model show that the solar irradiation at the Earth's surface depends to a much lesser extent on the absorption by aerosols, water vapour and ozone than on the cloud coverage.Thus to compute the global irradiation, these quantities may be specified by the use of climatological means. The radiation field of a cloud-free atmosphere is described by relations which incorporate seasonal and daily variations based on the radiative transfer calculations. Cloudiness is then parametrized by the optical thickness of the clouds and the cloud top heights. Since the latter factor is less important than the former, the cloud top height is estimated from simultaneous infrared data using climatological temperature profiles (Cole et al., 1965). So, for a given geographical area and for a given time there exists a unique relationship between the optical thickness of the cloud and the upward reflected radiation measured by the satellite which may permit estimation of the global irradiance on the ground.

The following quantities are defined:

G_{max} = maximum value of the global irradiance on the ground under cloudless conditions.

R_{min} = minimum value of the reflected irradiance under cloud-free conditions.

R_{max} = maximum value of the reflected irradiance above an optically thick cloud.

Then the following relations hold:

$$R_n = \frac{R - R_{min}}{R_{max} - R_{min}} \quad \text{, and} \tag{3.1}$$

$$G_n = \frac{G}{G_{max}} \tag{3.2}$$

where G is the time value of global irradiance

and the results of the radiative transfer calculations are expressed in the functional form:

$$G_n = f\ (R_n) \tag{3.3}$$

3.2.3 Utilisation of satellite data

After a navigation pre-process to correct the satellite imagery geometrically, spatial averages of 6x8 pixels (around 25x25 km) are formed. Since the satellite measures radiances L_R in uncalibrated units, a quantity analogous to R_n can be defined:

$$L_N = \frac{L_R - L_{min}}{L_{max} - L_{min}} \tag{3.4}$$

where L_{max} and L_{min} are the satellite-measured radiance equivalent of R_{max} and R_{min}. L_N is assumed proportional to R_n. This procedure avoids dealing with the calibration of the satellite data. Thus, one has only to determine L_N as an indicator of R_n from the data of the solar channel and the cloud top heights from the data of the infrared channel to estimate the global radiation G.

(a) minimum radiance maps

For the same local time (i.e. slot), minimum reflected radiance maps (weighted with cos) representing cloudless situations are derived for time periods of 15-30 days. Spatial filtering is needed to eliminate cloud shadows over areas of low sun, and corrections are to be applied over those areas where no cloud-free situations occurred. The use of the data of the IR-channel improves the results because cirrus clouds can then be more easily determined (see Fig. 2.2). This work may require some subjective analysis. It was found that in several regions over Europe a 30-day period complete by cloud-free sky did not occur. Then in addition a manual correction has been necessary assuming that the minimum radiance over those areas would be equal to those of the neighbouring cloud-free area. This assumption is useful as long as no strong spatial variations of ground albedo occur.

(b) maximum radiance function

Values of the maximum radiance L_{max} in the solar spectral ranges are required to obtain an upper limit of possible reflection from cloud surfaces where clouds are optically very thick. They are selected according to the criterion of almost homogeneous cloud surface as observable in the standard deviation of infrared radiances. Some diagrams are shown in Fig. 3.1

3.2.4 Comparisons with ground truth measurements

Comparisons with the European pyranometer network were performed for hourly and daily means of results which were determined for the months of June 1979 and April 1982. The June data were obtained with up to 6 slots per day while the April data were obtained with only 3 slots and less per day. Some tests were done to calibrate the physical parameters used in the model of radiative transfer against ground based-measurements.

The calibration of the model consists primarily in the correction for systematic deviations from simultaneous ground-based measurements. Here the authors selected specifically measurements at cloud-free conditions (fractional coverage smaller than 0.1). One further had to find whether all data show systematic offsets or not. Any corrections were primarily applied to changes in the model visibility (aerosol content) within the atmospheric boundary layer. While in June the visibility of 23 km led to excellent results one had to use a slightly higher visibility (smaller turbidity) for the April data.

The intercomparison between ground-based and satellite-data values of the global irradiance must take into account that these values are obtained from averages over different time spans. While the satellite measurements are made within fractions of a second over an area, the ground-based measurements have a temporal resolution of at least an hour or more. Therefore, one has to consider also the mean speed and temporal variability of cloud fields. This is demonstrated in Fig. 3.2, where the upper portion shows the effective area which must be chosen to compare with ground-based measurements. The lower part of this figure shows the temporal interpolation between two hourly averages of

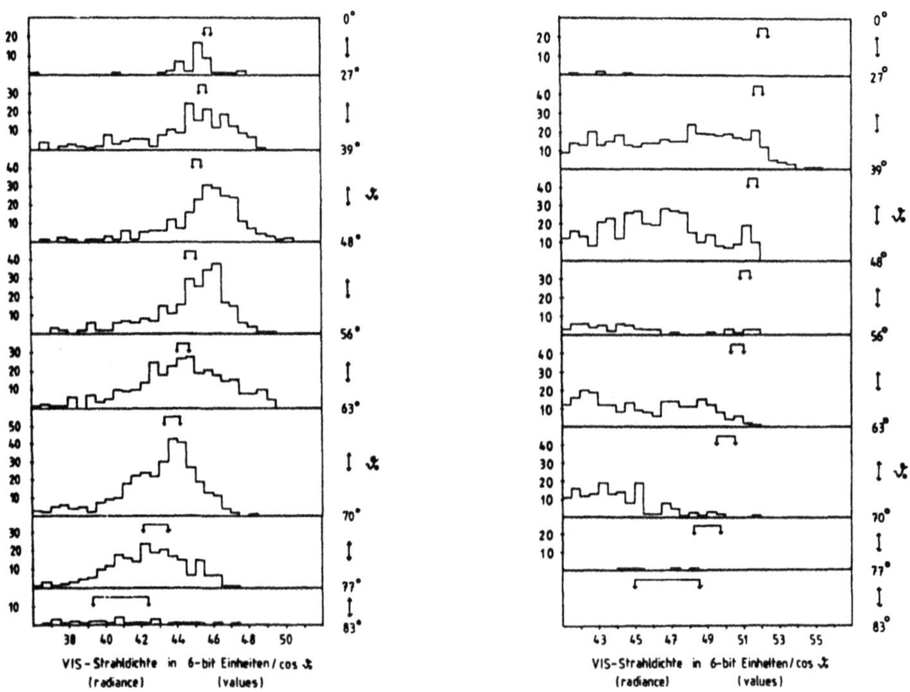

Fig. 3.1: Histograms of maximal radiances (divided by cos $_{o}$), measured above homogeneous cloud layers (height more than 13 km)
left: June 1979
right: April 1982

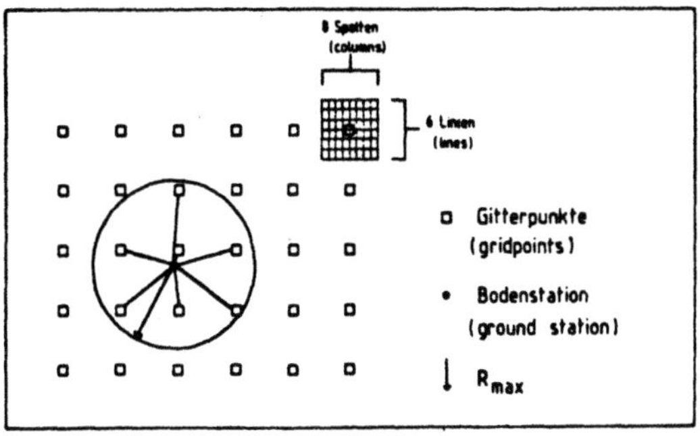

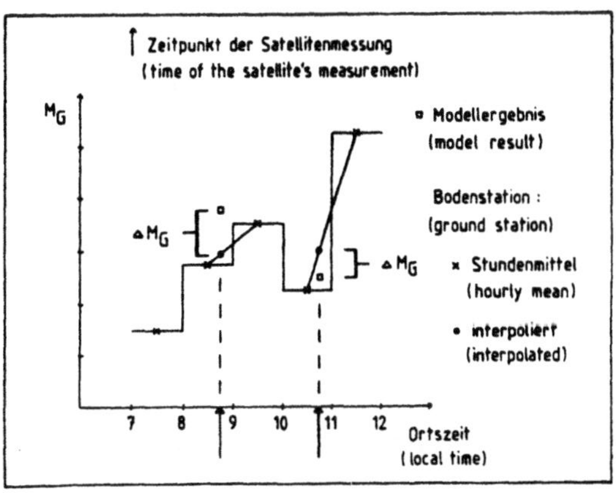

Fig. 3.2: **Top:** Areal coverage for intercomparison between ground-based measurements and satellite data.
below: Interpolation between two hourly means of pyranometer measurements

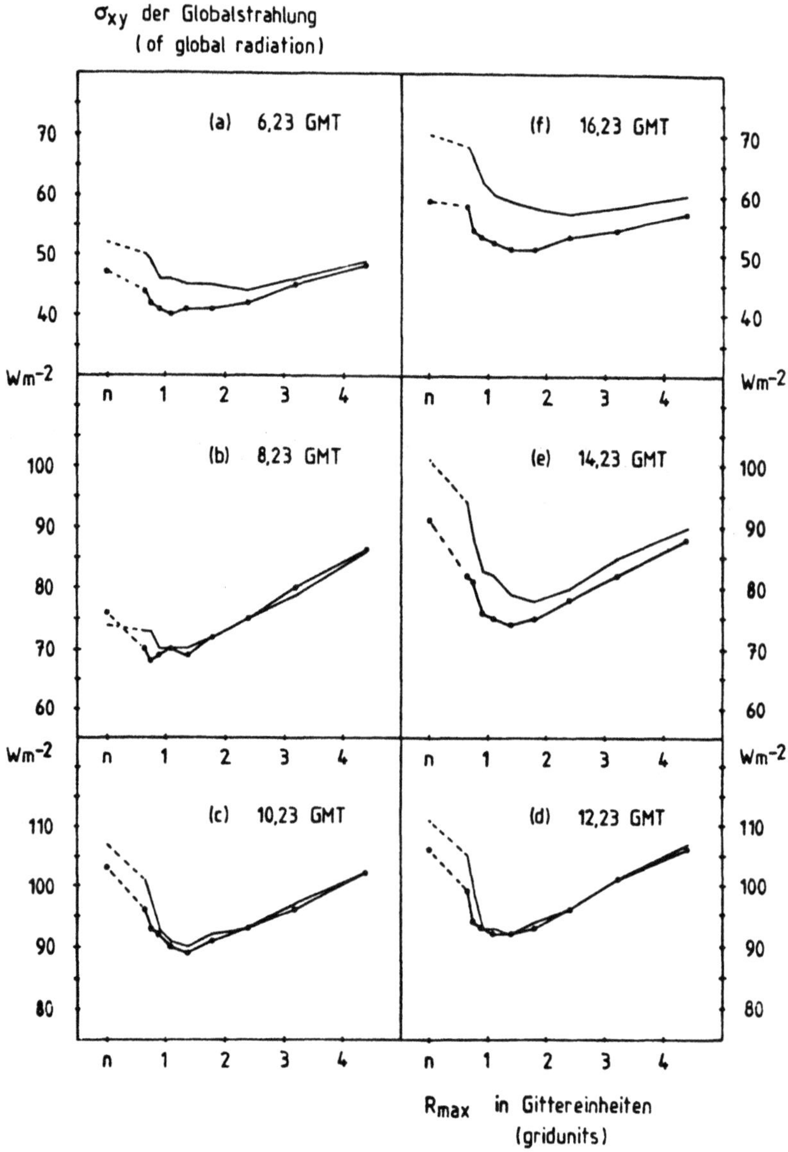

Fig. 3.3: Mean standard deviation of single values of the global radiation function of R_{max} (June 1979, Middle Europe)

ground-based measurements. The radius R_{max} depends primarily on the mean cloud speed. Further more, the determination of the spatial location of cloud shadows as viewed by the satellite must be considered. Therefore a geometrical correction is applied to take into account this possible source of errors.

As a universal average for all intercomparisons, a value of $R_{max} = 1.4$ has been used. Fig. 3.3 shows the effect of R_{max} on the mean deviation of satellite-data-derived values of the global radiation from ground-based measurements. Only measurements of about 23 stations of the German Weather Service were used here.

Daily sums and temporal averages are much more free of random noise and thus more accurate than the individual values, provided that enough measurements enter into them. We may study in Table 3.1 individual daily sums and their averages over 5 days and 30 days respectively, as obtained over different regions and for different times of the day. The table shows clearly that, with increasing number of days, there is a sharp drop in the standard deviation. The mean standard deviation of monthly averages falls to about only 6 to 7 % of the average which is close to the limit of acceptable accuracy of the ground-based measurements themselves. The same will also be found for daily sums (see Table 3.2).

	single day average		five day average		monthly average		
	in Wm^{-2}	in % of the average	in Wm^{-2}	in % of the average	in Wm^{-2}	in % of the average	
Fed. Republic of Germany (only)							June 1979
	41	18.5	21	10.2	10	4.7	6.23 GMT
	69	16.5	43	10.6	31	7.3	8.23 GMT
	89	17.5	50	9.9	25	4.8	10.23 GMT
	92	18.7	53	10.7	30	6.1	12.23 GMT
	74	18.4	42	10.3	28	7.0	14.23 GMT
	51	21.7	29	12.3	18	7.8	16.23 GMT
							April 1982
DK	50	11.8	29	6.9	16	3.9	8.53 GMT
EIR	51	16.8	29	8.8	18	5.9	
GB	49	15.0	27	7.6	14	4.4	
D	57	13.0	33	7.2	20	4.5	
F	56	11.5	37	7.4	27	5.5	
DK	70	13.4	38	7.2	22	4.2	11.53 GMT
EIR	93	20.2	50	10.3	34	7.4	
GB	90	18.4	53	10.4	33	6.7	
D	84	16.3	51	10.1	35	6.9	
F	87	13.2	66	9.7	52	7.9	
DK	50	16.1	35	11.4	25	7.9	14.53 GMT
EIR	55	14.2	34	8.7	24	6.1	
GB	66	17.1	38	9.6	29	7.5	
D	60	19.3	36	11.6	22	7.1	
F	64	14.2	43	9.5	33	7.3	

Table 3.1: Mean quadratic deviation of the intercomparison of single values of the global radiation for three different intervals of averaging (1, 5, 30 days) and for different times of the day as indicated in the right-hand column.

	single day average		five-day average		monthly average		
	in $kWhm^{-2}$	in % of the average	in $kWhm^{-2}$	in % of the average	in $kWhm^{-2}$	in % of the average	
							June 1979
Eastern Europe	0.61	10.5	0.41	7.2	0.28	4.7	up to
Middle Europe	0.57	11.8	0.35	7.3	0.21	4.3	6 slots
Western Europe	0.55	10.2	0.33	6.1	0.25	4.7	per day
							April 1982
DK	0.50	11.9	0.26	6.0	0.14	3.3	up to
EIR	0.52	13.7	0.35	8.7	0.25	6.7	3 slots
GB	0.55	14.0	0.32	7.6	0.23	5.7	per day
D	0.49	11.5	0.29	6.9	0.22	5.1	
F	0.57	11.0	0.42	7.7	0.35	6.8	

Table 3.2: Mean quadratic deviation of the intercomparison of daily sums of the global radiation for three different intervals of averaging (1, 5, 30 days).

The June data were obtained with up to 6 slots per day and thus appear more accurate than the April data which were obtained with only up to three slots per day.

The monthly averages are finally compared in Fig. 3.4 for both data sets (June 1979 and April 1982). The standard error of estimate amounts here to 0.25 and 0.28 $kWhm^{-2}$, which is less than 5% of the average.

The three plots of Fig. 3.5 compare monthly averages of the values derived for each slot (hourly sums) for only 78 stations of the countries indicated in each figure. A somewhat larger scattering occurs if only daily sums are compared with the simultaneous pyranometer data. This is shown in Fig. 3.5 where the standard error of estimate is now 0.56 $kWhm^{-2}$, corresponding to about 10% of the average.

3.2.5 Maps of global radiation

Fig. 3.6 shows the results obtained from the early morning (8.23 GMT) and early afternoon (14.23 GMT) slots of 17 June 1979. These results show clearly the effect of clouds which were present over Europe. On the left and right hand side one also observes the strong variation of the Sun's zenith angle.

Fig. 3.7 compares the daily sum of the global irradiance for 17 June 1979 which have been derived from data of 6 slots. The lower part of Fig. 3.7 shows the five-day average of the daily sum for the period from 16 June to 20 June 1979. One sees clearly that the average over these five days already causes a smoothing of the contours. Major cloud fields have been observed over southern Germany and Austria and over adjacent eastern countries.

Maps of the monthly averages of hourly sums obtained for two slots (8.23 GMT and 14.23 GMT) are shown in Fig. 3.8. There we see clearly the effect of convective cloudiness in the afternoon data over mountainous regions; while in the morning these areas appear very clearly. Fig. 3.9 shows monthly averages and their standard deviations of daily sums for June 1979, and Fig. 3.10 shows the same for the second data set (April 1982). The grey scale has been chosen in intervals such that interesting structures are enhanced. It can be seen clearly that convective activity in June considerably enhances the differences between land and ocean. The June image also shows strong variations of the global radiation over regions with orographically induced cloudiness, while in April stong variations occur over the Mediterranean due to travelling disturbances.

The European network of pyranometers and sunshine recorders is not dense enough to provide such high spatial resolution maps of the global radiation. Moreover, there are no measurements available over the oceans and over desert regions. Furthermore, the measurements over individual countries may deviate considerably up to 8% of a reference value (Plazy et al., 1982) although they should be more accurate (about 2%, WMO, 1971). This is due to different techniques of data handling, maintenance of the instrumentation and in particular maintenance of the reference standards.

For our specific months there is no map available of global irradiance which has been derived from pyranometer data. However, Kasten et al. (1982) collected all data from a ten-year period (1966-1975) which could be derived from either pyranometer or sunshine recorder measurements. In these maps one sees also patterns which can be observed in Fig. 3.9. For instance, the global radiation near the coastline of northern Spain, over Italy and Yugoslavia is much smaller than in the neighbourhood. But the resolution is much lower and not so useful for application to the planning of solar energy plants.

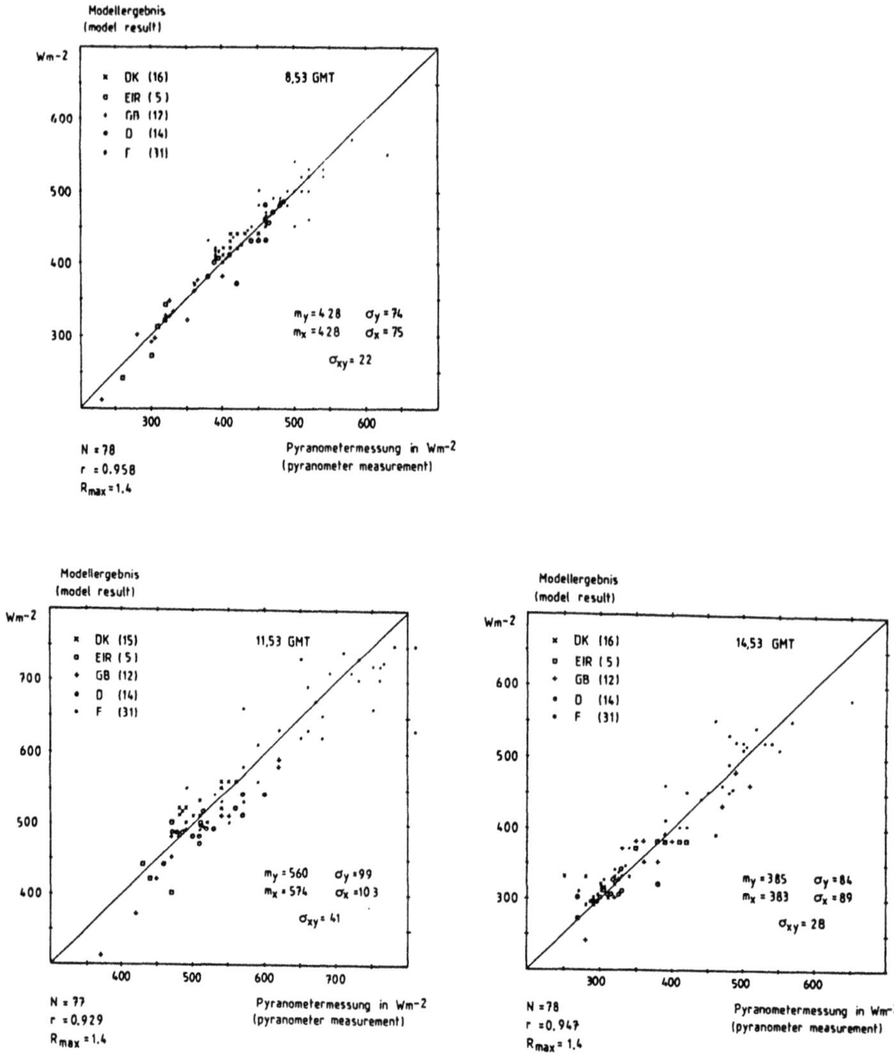

Fig. 3.5: Intercomparision between satellite-derived and ground-based measurements of the global radiation (hourly sums)

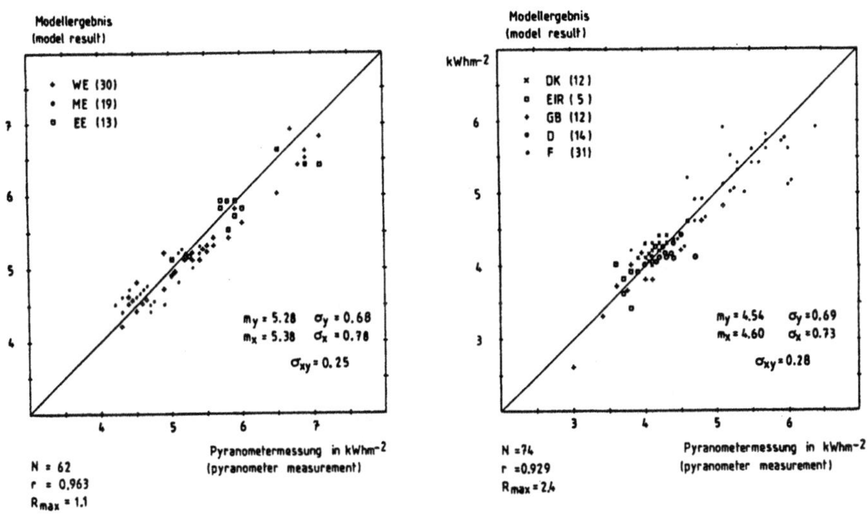

Fig. 3.4 : Comparision of monthly averages of daily sums
left: June 1979
right: April 1982
n = number of stations
r = correlation coefficient
R_{max} = maximal radius of distance around a station

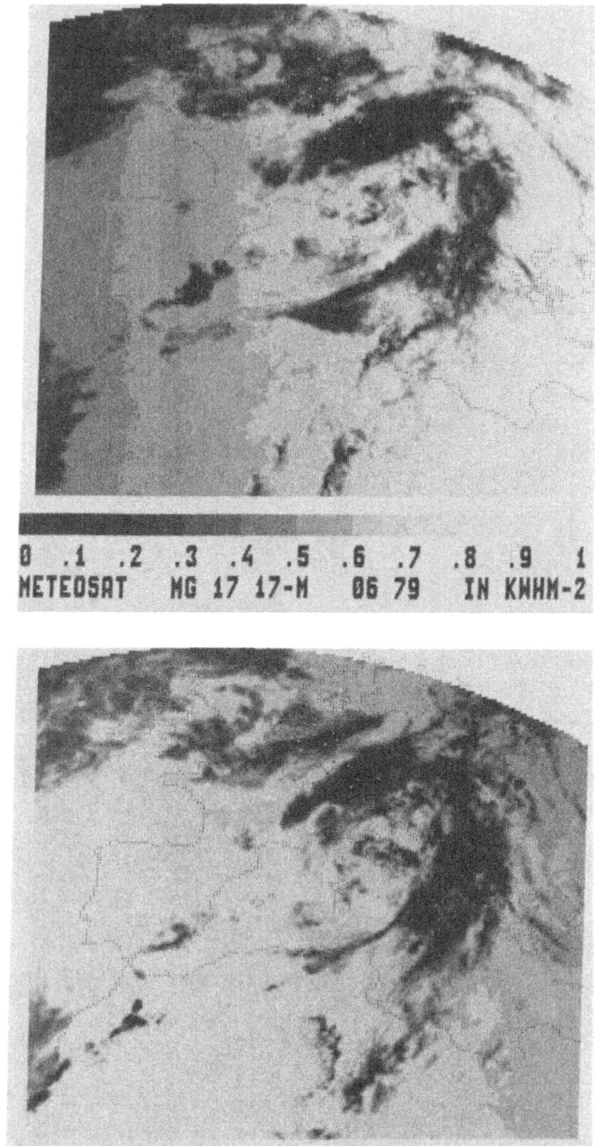

Fig. 3.6: Maps of the global radiation (in kWhm-^{2})
Top: hourly sums (8.23 GMT) for 17 June 1979
Bottom: hourly sums (14.23 GMT)

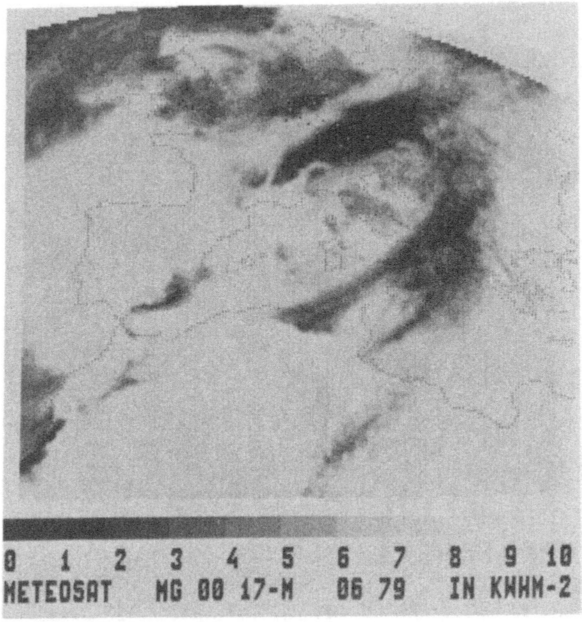

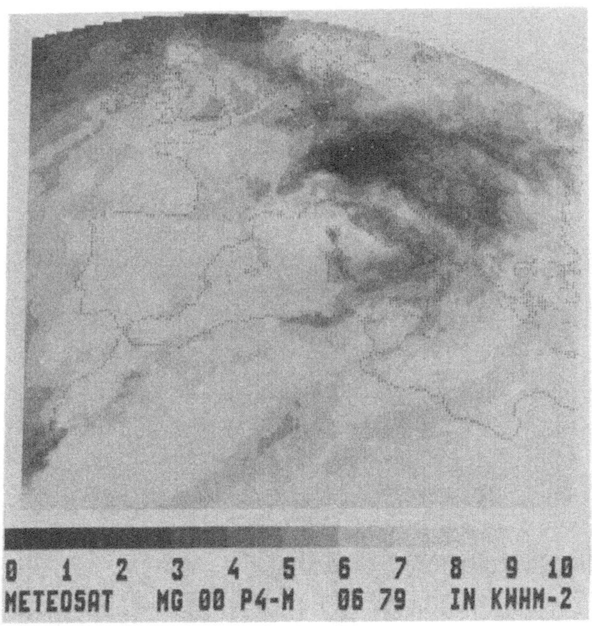

Fig. 3.7: Maps of the global radiation (in $kWhm^{-2}$)
Top: daily sum for 17 June 1979
Bottom: 5-day average of daily sum (16-20 June 1979)

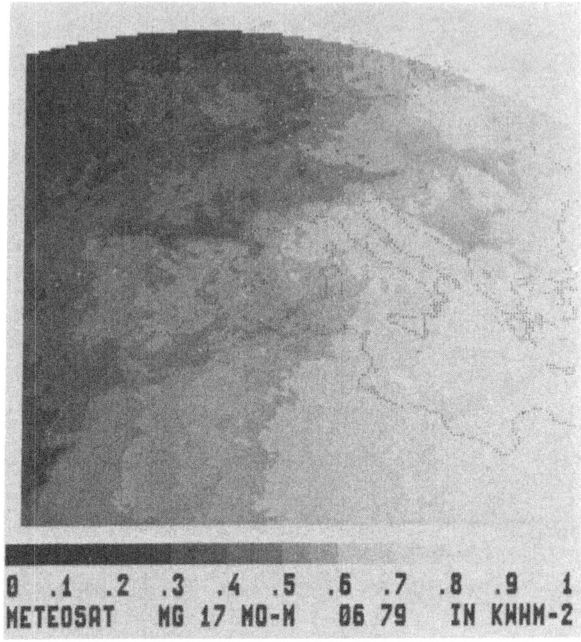

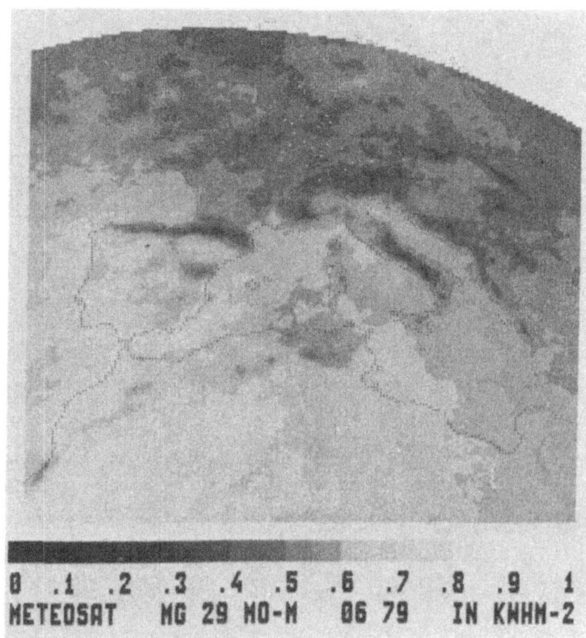

Fig. 3.8 : Maps of the global radiation (in $kWhm^{-2}$)
Top: monthly average of hourly sums (8.23 GMT) for June 1979 (23 days);
Bottom: monthly average of hourly sums (14.23 GMT) for June 1979 (28 days)

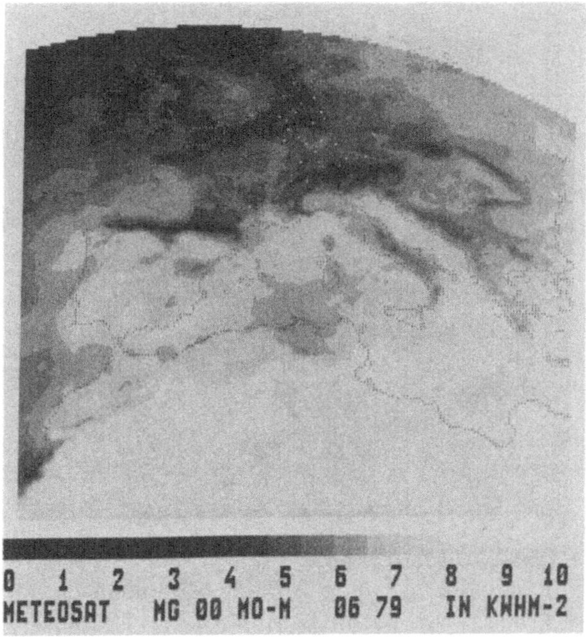

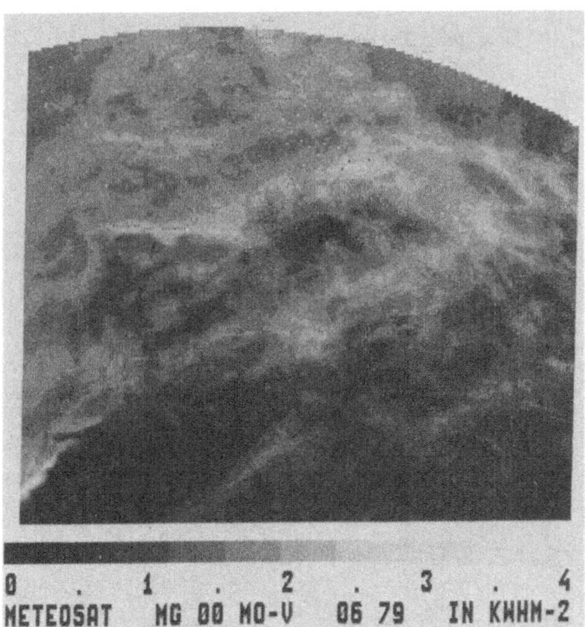

Fig. 3. 9 : Maps of the global radiation (in $kWhm^{-2}$)
Top: monthly average of daily sums (1230, June 1979)
Bottom: standard deviation of daily sums (1230, June 1979)

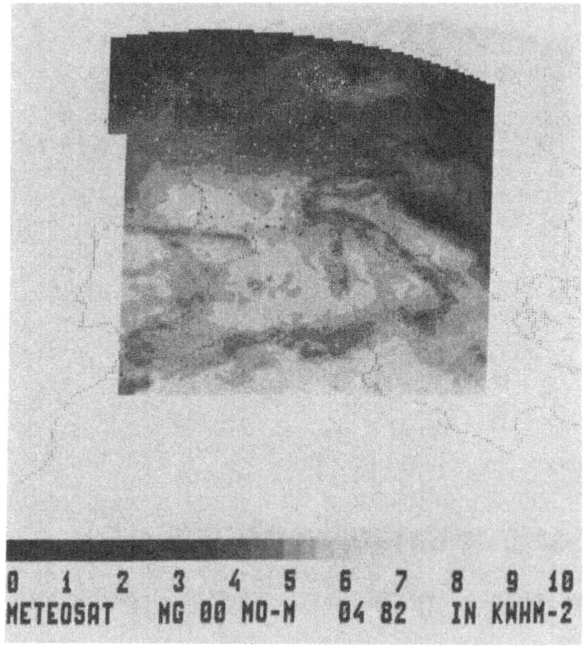

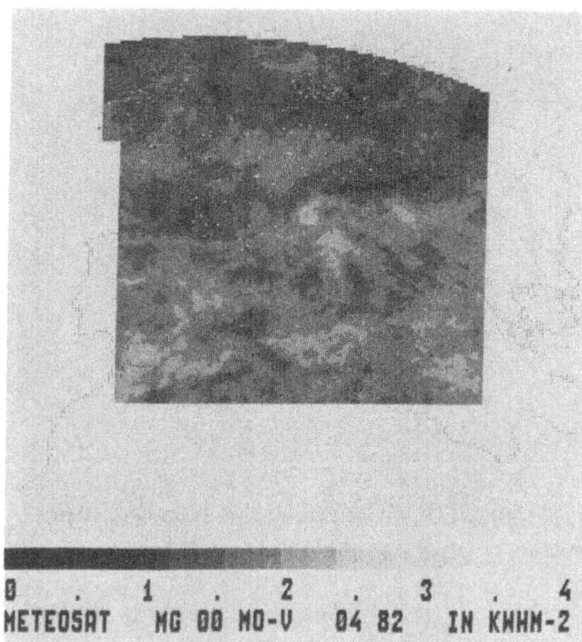

Fig. 3.10: Maps of the global radiation (in $kWhm^{-2}$)
Top: monthly average of daily sums (1230, April 1982)
Bottom: standard deviation of daily sums (1230, April 1982)

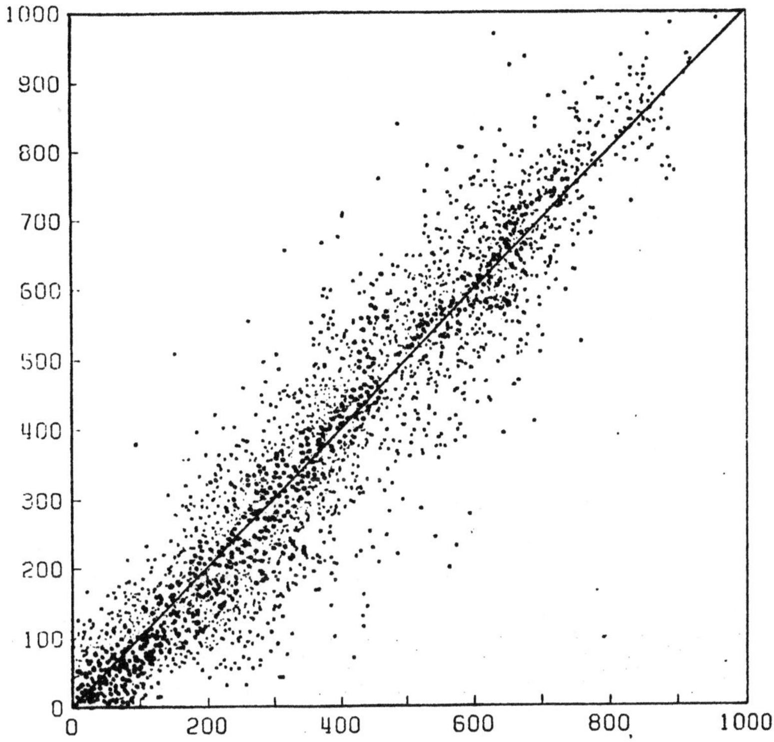

Fig. 3.11: Correlation of 3 hour intervals; Stuttgart

(Daily sums in Jm^{-2})

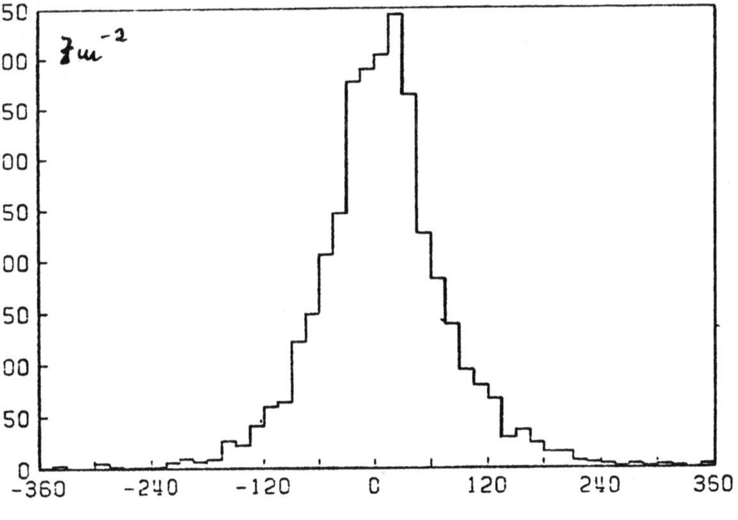

Fig. 3.12: Error histogram for 3 hour intervals; Stuttgart

3.3 The method of Reinhardt et al. (Stuttgart)

The method used at Stuttgart is a purely statistical one. It has been developed by R.E. Reinhardt and, P. Schwarzmann, of the Institute for Physical Electronics, University of Stuttgart.

3.3.1 Introduction

The method is based on two assumptions:

- Values of the global irradiation on the Earth's surface are completely defined by the atmosphere.
- In principle, a deterministic relation exists between spectral intensity measured by satellite sensors and global irradiance incident on the Earth.

The success of processes to estimate global irradiation from satellite images depends on a sufficiently unambiguous relation between the global irradiation and the spectral intensity of satellite images. Bearing in mind the enormous number of parameters characterizing the atmospheric state in detail with respect to radiation transfer and the small field of view of satellite sensors combined with the restriction to measure only remote sensable values of the atmosphere, the presupposition cited above is well legitimated.

Therefore the situation outlined above favours the application of statistical methods characterized by maximum freedom in presuppositions.
Significant information for the estimation of global irradiation can be extracted from spectral satellite images. The main tasks to be solved are:

- Determination of relevant image information (features x_i) which correlate with the global irradiation.
- Determination of estimation function $f(x_i)$ on the basis of ground truth data of a training set.

As regards the large amount of information presented by spectral satellite images, statistical methods seem to be more powerful than intuitive approaches.
The basic ideas for the introduction of a statistical approach to estimate global irradiation from satellite images are summarized below:

- Statistical methods make it possible to introduce in the algorithms large numbers of features derived from the available measurements (satellite images) - hence all available information may be included in the process.

- The methods include a selection process in which all non-relevant features may be excluded to further compress the data to be processed.
- A maximum of freedom is provided to extract features of the primary data with the possibility to apply highly computer-oriented fast feature extraction processes based on image-processing techniques.
- A priori information is easily introduced into the process by coding the information into features. Such a priori information may be for instance:

  meteorological rules gained by experience
  selected physical rules
  geographical features (e.g. land, sea etc.)
  features which are obvious by extremely relevant for global irradiation (e.g. clouds)
- Local neighbourhood relations important for radiation integrals over time intervals can be introduced.
- Data bases containing labelled training sets (feature sets from the satellite image plus ground truth data) permit the application of self-training and optimizing strategies which are common and proved in classification theory.
- The problem is divided into an expansive training phase and a very effective run phase with the possibility to adapt the classifier further during the run phase by increasing the data bases with sporadic ground truth data.
- By using all the information in the satellite images the introduction of albedo information can be avoided.
- The strategy is easily adapted to other image material (other satellites) by simply rerunning the training process with a labelled data base containing the feature sets of the new image material.

Based on the ideas cited above, a method was developed to estimate the global irradiation for different time intervals and daily sums. For the experiments, NOAA imagery of a complete month (April 1982) was used.
The following section describes in detail the method developed and information extracted from the satellite images.

3.3.2 Strategy and algorithms to estimate global irradiation

Learning methods are strongly dependent on the labelled data sets provided to the algorithm in the training phase. Fundamentally important is the statistical relevance of the training set; in other words: the algorithm in the training phase must have seen all situations which it will have to process in the run phase. Therefore the training set has to be prepared carefully.

Satellite images record the momentary situation whereas available ground truth data of global irradiation are delivered as hourly sums. This difference is ofno consequence if the state of the atmosphere is homogeneous. But in partly cloudy situations the hourly variations of the global irradiation may be large. This problem is overcome by the following assumption: Short time variations of global irradiance are primarily due to the movement of the weather situation and less to locally fixed variations. Therefore temporal variations may be substituted by spatial variations.

Relation between geographical and image coordinates:

To combine ground truth measurements of global irradiation and radiance calculated from satellite images, one must know the equations to transform geographical coordinates to satellite coordinates. These equations are derived from landmarks determined by visual inspection of the satellite images and from the location in geographic maps. Then a computer programme calculates the x,y coordinates in the satellite images of all ground-truth stations. The last step is the rectification process. This step represents a dilatation in the x-direction. Information is introduced into the generated gaps by interpolation in the sense of the nearest neighbour approximation. The result of the preprocessing steps is a set of 5 (for NOAA 7 scenes, and 4 for each NOAA 6 scene) rectified subimages 64 x 64 pixels for each ground station (one subimage per satellite spectral channel).

NOAA-satellite images of the areas of interest must be available 3 times per day (morning, afternoon, evening). If fewer images are available the estimations become less accurate; if more are available (due to overlapping of consecutive orbits) estimates become more accurate. To cover all countries of the European Community, data from at least two consecutive orbits are needed.

The application of all later steps requires some preprocessing steps of the satellite images which have already been described:

- match of satellite image to geographical coordinates;
- rectification of subimages.

By visual inspection of images, cloudy and clear areas have been detected. The features which characterize pixels as belonging to a cloudy or clear area are their grey value in the different spectral images. Each pixel defines a point in a 5 dimensional (for NOAA 7 scenes, and a 4 dimensional for NOAA 6 scenes) features space the axis of which are represented by the magnitudes of the spectral grey levels. Standard classification strategies can be applied to separate optimally the two point ensembles (cloudy-clear). In this case a linear classifier separating the two classes was applied. The separating criterion has the form:

$$y_1 = a_o + a_1x_1 + a_2x_2 + a_3x_3 + a_4x_4 + (a_5x_5) \tag{3.5}$$

(estimation: to be a cloudy pixel)

$$y_2 = b_o + b_1x_1 + b_2x_2 + b_3x_3 + b_4x_4 + (b_5x_5) \tag{3.6}$$

(estimation: to be a clear pixel).

The coefficients a_n, b_n are determined by the classifier in the training phase. The parameters a_o, ..., a_n and b_o ,......., b_n are stored for use in the operational phase of the process.

By applying the classification formula, all pixels of the rectified subimages are classified as cloudy or clear pixels. The result is a binary image describing the cloud situation of the subimage. This binary cloud image is added to the subimages data set as a 6th (or 5th) subimage.

A feature set describing each subimage set is generated. To miss as little information as possible, the feature set includes 360 values at the beginning and comprises two groups of features:

- Features derived from the subimage set by image processing methods
- Features derived from a-priori knowledge

The complete feature set is listed in Table 3.3.

The feature group comprising the a priori information includes the numbers 335-344 and needs no further interpretation. The other feature group is derived from the subimages by processing the images.

Table 3.3: Complete feature set

Feature Number	Description
1 - 100	Area: 50 measurements for absolute grey values and 50 measurements for relative grey values.
101 - 200	Circumference: 50 measurements for absolute and relative grey values.
201 - 250	Connexity: 25 measurements for absolute and relative grey values.
251 - 260	Minima and maxima of the 5 subimages.
261 - 270	Measurements in binary cloud image: area, circumference, connexity in the eroded and dilatated cloud image; Cloud area of the 5x5 pixel central image.
271 - 295	Grey value of the centre point in median filtered images.
301 - 320	Grey value ratios of different spectral images (central point)
335	Angle sun, satellite
336	sin (zenith, satellite)
337	sin (elevation of the Sun)
338 - 341	Clear sky radiation sums
342 - 344	Weights of flights

Area, circumference and connexity are defined as Minkowski values.

Statistical rules require that the number of features must be considerably lower than the number of cases from which they are derived. Thus a feature selection procedure has to be applied. The criterion elected for this selection is the correlation between the feature and the ground truth values of the hourly global irradiation in 3 time intervals of the satellite.

For further processing for each time interval the 25 most correlated features have been selected.

In a multidimensional (25) correlation experiment, the correlation between the complete feature set and the global irradiation is determined. The results of such a multidimensional correlation experiment are the correlation constant R, the standard deviation δ, the parameters c_o, c_1, ..., c_{25} for an estimation equation for the global irradiation. The parameter set c_o, c_1, ..., c_{25} is stored as the only data needed for use in the later run phase.

The estimation of the daily global irradiation will be based upon:
- 1, 3, 5 hour interval values of hourly global irradiation
- Physical radiation model without atmosphere
- Statistical approach which combines model parameters and interval values.

Physical model:

Let us define the radiation.

$$g_n = \frac{F_n}{G_c} \tag{3.7}$$

with F_n the hourly irradiation in the clear sky model for 1, 3, 5 hour intervals centred at the flight time and G_c is daily sun of clear sky irradiation. g_n describes then that part of daily irradiation in the clear sky model which has been calculated from the image of one flight, and is called the "model weight".

Based upon a real measurement M_n for the interval Δt an estimation of daily sum can be provided by

$$E_{mn} = \frac{M_n}{g_n} \tag{3.8}$$

If more than one flight is available per day, the weighted estimations of the different flights are added.

$$E_{mn} = (M_{n1} + M_{n2} + \ldots + M_{nk}) \frac{1}{(g_{n1} + g_{n2} + \ldots + g_{nk})} \tag{3.9}$$

This procedure is performed separately for the 1, 3 and 5 hour estimates of the interval sums. Therefore 3 daily sums derived from the physical model are available (E_{m1}, E_{m3}, E_{m5}) and are transmitted to the last step as the variables x_2, x_3, x_4.

Statistical model:

In a first step a regression analysis is performed to determine the correlation coefficients between the subimage features and the daily global irradiation. The 25 best coefficients have been selected for further processing. With these coefficients from the training phase, daily sums for each flight can be estimated (E_s). The values of the different flights of the same day are added by weighting the single flight values with the model weights.

This procedure is performed for the weights of the 1, 3 and 5 hour intervals. A furth daily sum is estimated by simply adding all daily sums (E_s). The results of this model are therefore 4 daily sums which are transmitted to the last step as the variables x_8, x_9, x_{10}, x_{11}.

Further variables to be transmitted to the last correlation block are the clear sky maximum global irradiation G_o as the variable x_1 and the sums of the weights of all flights of a day ($G_n = \Sigma g_1$) separate for the 1, 3 and 5 hour values as the variables x_5, x_6, x_7.

All the steps described deliver 11 values to the final regression step. This step performs a regression analysis with respect to the daily sum. The 8 best correlation coefficients are stored to be used in the run phase to estimate the daily global irradiation.

3.3.3 Results

The results will be presented and discussed in the following order:

- Primary data material
- Cloud detection
- 1, 3, 5 hourly interval global irradiation
- Daily global irradiation

All experiments have been performed with NOAA 6 and NOAA 7 material from April 1982. All satellite passes used have been received by DFVLR - Oberpfaffenhofen or by Dundee University. The primary data are stored on Computer Compatible Tapes (CCT's). The problem of the compatibility of hourly global radiation with instantaneous intensities of satellite data has been discussed earlier. To adapt the hourly sums recorded for full hours to the flight times, the global irradiation data have been interpolated between the nearest neighbours.

The cloud detection classifier was trained with 7500 labelled pixels. The classification result for the training set shows that this binary classifier is not efficient and tends to classify several clouds as clear areas. This inefficiency is explained by the broad intermediate range between very thick clouds and clear skies. It further reflects the effect of the surface albedo to upward radiances.

Global radiation intervals of 1, 3, 5 hours centred round the flight time have been calculated for all stations for which satellite image material and ground truth data have been available. To compare ground truth values and calculated values, the correlation parameters and the standard deviation of these values have been calculated. The results are summarized in table 3.4

It is surprising that the 3 hour values achieve the highest correlation coefficients, followed by the 5 hour values. It is believed that the substitution of the temporal variations of the global irradiation by the spatial variations of the global irradiation with respect to a 70 x 70 km surrounding of the ground stations is best for 3 hour sums and worst for 1 hour sums.

Table 3.4 Estimates of daily sums of global irradiation (Jcm^{-2} for time intervals of 1, 3 and 5 hours.

Interval (hours)	All stations			Germany - Gr. Britain		
	1	3	5	1	3	5
Cases	3884	3884	3884	1241	1241	1241
Correl.	0.95	0.96	0.95	0.95	0.96	0.96
Standard Dev.	27.5	70.8	111.4	24.7	63	98.9

In Fig. 3.11 are compared the calculated (x axis) and ground truth values (y axis) for the 3 hour experiment of all stations. In Fig. 3.12 the error histogram (calculated minus ground truth value) is presented for the same experiment.

Daily sums have been calculated for all ground stations and days for which at least one flight per day was available. The results are presented in the form of correlation coefficients and standard deviations (see Table 3.5).

Table 3.5 Daily global irradiation (Jcm^{-2})

Flights/day	1	2	3	4
Cases	1175	927	467	166
Correlation	0.87	0.90	0.94	0.97
Standard Dev.	265	235	202	157

As the number of flights per day increases, the correlation improves and the standard deviation of the estimated results decreases. This is also illustrated in the histograms shown in Fig. 3.13.

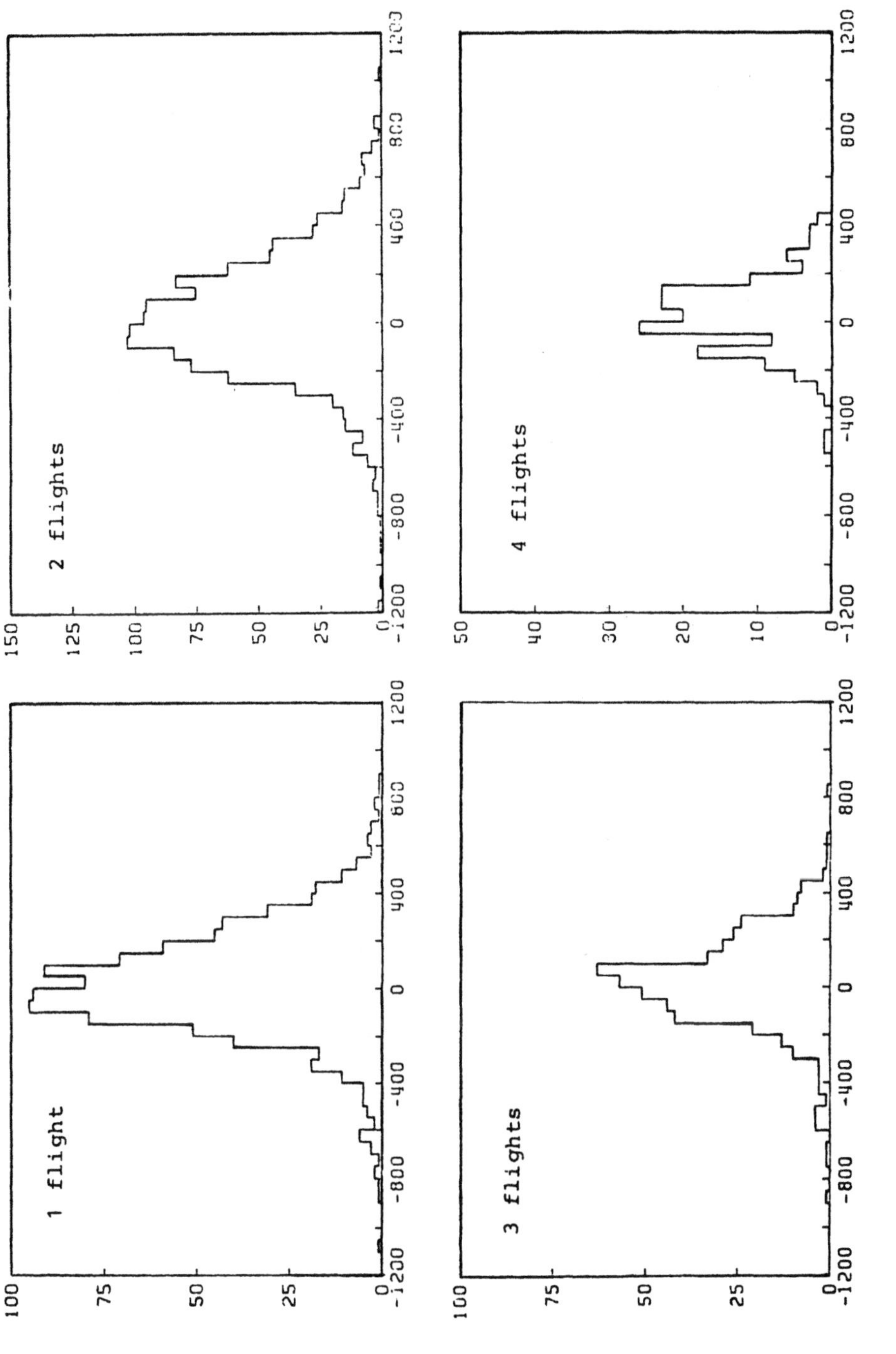

Fig. 3.13: Error histograms of daily sums of global radiance, calculated from data of 1 to 4 flights per day (Jm^{-2})

3.3.4 Conclusions

The results presented here demonstrate a high correlation between the measured ground truth values and the estimated values. If statistical requirements concerning the training set are met and the problem of short-time measurement of the satellite image and hourly sum of ground truth data is borne in mind, the accuracy of the estimation is sufficient for designing and operating solar energy installations.

The procedure is adaptable to other satellite images by rerunning the training algorithm with a new labelled data set. Albedo information is not required.

Despite its mathematical elegance this method did not consider the temporal sampling problem since it used only data of the NOAA polar orbiting satellites. This may be possible if applied to the temporally more dense data of geostationary satellites.

3.4 Method of Monget et al. (Sophia-Antipolis)

D. Cano and J.M. Monget developed the method used at Sophia-Antipolis which is a statistical approach for the estimation of global irradiation using satellite data.

3.4.1 Introduction

The basic idea of this approach is that the amount of the cloud cover over a given area determines the global irradiance at that area. Thus the method is divided in two steps. For each pixel (i,j) of the satellite image, one first computes a cloud cover index which is then used as a basis for a statistical estimation of the global radiation. A preparatory step, the construction of a reference map of ground albedo, is also needed and is done at the very beginning of the process. Afterwards this ground albedo map is updated daily. The cloud cover index map is derived from the comparison of the current satellite image and the ground albedo map. Following this determination, a statistical regression between pyranometer data and the cloud cover index at the same locations is applied. Finally the global radiation map is computed using the interpolation technique between the ground station locations.

The data used by the group at Sophia-Antipolis are from the geostationary satellite METEOSAT, measured simultaneously in the solar as well as in the infrared range. Each image is pre-processed in the following way: first it is geometrically corrected using landmark locations and correlation techniques, then the noise is removed by filtering and finally numerical counts in the solar image are normalized by dividing the count of each pixel by the direct spectral irradiance which would be measured on a horizontal plane located at this pixel in the absence of the atmosphere.

Because the numerical count can be directly linked to the upward radiance, this normalization is equivalent to the computation of a reflectance factor. Furthermore, since most continental surfaces are similar to a Lambertian reflector for the angles concerned (except water and snowy surfaces), it is also equivalent to the computation of an image of the albedo of the ground more or less covered by cloud. In fact, the exact relationship between the upward radiance and numerical count is unknown because the METEOSAT satellite has no calibration. For convenience we will use the term albedo.

3.4.2 Construction of the ground albedo map

Except over snowfields and deserts, the cloud albedo is higher than the ground albedo. Usually one makes use of this fact in the construction of the ground albedo reference map by assuming that, given a satellite image time series, the clouds are non-stationary and by taking in each pixel the minimum of the values observed this area. This minima map is then called the ground albedo reference map (see also the method by Möser and Raschke, chapter 3.2).

If the cloud albedo is greater than the ground albedo, then in a time series of satellite images, the appearance of a cloud in the field of view of the sensor results in a jump of the measured radiance. It follows that the presence of a cloud may be detected by the difference between its high response and the response for a clear sky. The global irradiance for clear skies is computed with the relation:

$$G_c(t) = I_c(t)\ (\sin (h))^{(1+a)} \tag{3.10}$$

where G_c = global irradiance at ground level

$I_c(t)$ = solar irradiance at a plane outside the atmosphere, normal to the Sun's rays, as a function of time t

h = elevation of the Sun

The parameter $\underline{a}$ characterizes the total atmospheric transmittance and has been found to range between 0.14 and 0.17 depending upon the geographical location. In this work a value of 0.15 has been chosen and therefore the irradiance measured at satellite altitude $G_r(t)$ is expressed:

$$G_r(t) = \rho(t)\ I_c(t)\ \sin (h)^{1.15} \tag{3.11}$$

which is the irradiance measured by the sensor in the absence of clouds and coming from the ground albedo $\rho(t)$.

Then, given a time series of satellite images, it is assumed that the clouds are not stationary, that is, for each pixel there exists at least one occasion during which the clear sky model holds. The approximate character of the clear sky model and the instrumental noise are taken into account by considering that the variables are random. Finally the ground albedo is evaluated in each pixel in a recursive fashion by minimizing the variance of the difference between the measured radiances and the radiances resulting from the clear sky model, the cloudy cases being eliminated at each step.

The method is illustrated in Fig. 3.14 displaying the ground albedo map of Europe obtained from ten METEOSAT images of May 1979.

Fig. 3.14: Ground albedo map of Europe obtained from METEOSAT data for the period 5-21 May 1979
No correction has been made for cloud contamination.

Of course if a cloud layer is stationary over a location in the ten images, the initial hypothesis is not valid and the apparent albedo is not the ground albedo but the albedo of the cloud. This is the case over Ireland in Fig. 3.14.

It has furthermore been assumed that:

- the ground albedo is a stationary random variable ρ^S of mean $\bar{\rho}^S$ and variance σ_S^2 ,
- the cloud albedo is a stationary random variable ρ^C of mean $\bar{\rho}^C$ and variance σ_C^2 ,
- the apparent albedo is a stationary random variable x equal to ρ^S with a probability p^S and equal to ρ^C with a probability p^C.

If the difference between the averages $\bar{\rho}^C$ and $\bar{\rho}^S$ is much larger than the standard deviation σ and σ_C ,

$$\bar{\rho}^S - \bar{\rho}^C >> \sigma_S, \sigma_C \qquad (3.12)\ (3.13)$$

then the iterative minimization of the variance of the difference between the observed radiances and the ones issued from the clear sky model is equivalent to an iterative computation of the mean $\bar{X}$ of the variable X, when the cloudy pixels are eliminated in each step.

$$\bar{X} = p^S \cdot \rho + p^C \cdot \bar{\rho}^C \qquad (3.14)$$

This iterative average converges to the ground albedo value $\bar{\rho}^S$.

3.4.3 Evaluation of the cloud cover index

A cloud cover index (n^t_{ij}) is defined at a given time t, as a function of the characteristic albedo of the earth (ρ^S_{ij}), the apparent albedo at the same point as measured by the satellite (ρ^t_{ij}) and the average albedo of the cloud tops ($\bar{\rho}^C$). It is computed for each pixel using the following formula:

$$n^t_{ij} = (\rho^t_{ij} - \rho^S_{ij}) / (\rho^C - \rho^S_{ij}) \qquad (3.15)$$

The ground albedo of the Europe area was determined in the previous step for each 5x7 km pixel using an iterative algorithm for the detection and elimination of cloudy areas. The computation of cloud albedo (^{C}) is performed using an inverse algorithm retaining all the cloudy areas followed by the construction of an albedo histogram, which modal value is retained as an estimation of ρ^C.

The cloud cover index ranges from 0 to 1% and can be interpreted as the percentage of the surface of the pixel covered by clouds. On the other hand, since the transmittance of the clouds is highly variable, it can also be seen as an indicator of the transmission of an atmospheric column above a pixel, a low index denoting a high transmission factor.

The above method does not perform accurately when the albedo of the surface is of the same order of magnitude as the albedo of the clouds. In Europe this is specifically the case of snow-covered areas. In this case, an alternate cloud index is defined using the radiance measured by the satellite in the thermal infrared:

$$n'^t_{ij} = (L^t_{ij} - L^S) / (L^C - L^S) \qquad (3.16)$$

In this formula, L^C is the radiance of the clouds, and L^S is the radiance of cloud-free snow areas, both determined using image histograms in the thermal infrared. L^t is the measured "mixed" radiance.

The reference map of the ground albedo of Europe was computed using the METEOSAT images for the period 5-21 May 1979. However, long-term seasonal variations affect the albedo map. In order to take these variations into account, the albedo map is updated in the following way. During the processing each image is compared to the ground albedo map, pixel per pixel. If the pixel is not cloudy, its value is the instantaneous ground albedo value. This value is averaged with the previous one, the latter having more weight in the average than the former in order to avoid noise.

3.4.4 Statistical relationship between cloud cover index and ground-measured transmission factor

Contrary to the methodology used by other authors (Gautier, 1979; Raschke and Preuss, 1979; and Amado, et. al., 1981, who are using physical models of the transmission through the atmosphere this method employs a purely statistical approach to estimate the hourly global irradiation from satellite measurements.It is based on a linear relationship between the satellite-determined cloud cover index and the ground-measured transmission coefficent. This simple and efficient approach is based on the work of the Centre d'Energêtique de l'Ecole Nationale Supérieure des Mines de Paris (Bourges, 1979) and correlations determined by Pastre (1981) between duration of sunshine insolation and the cloud cover index.

The total atmospheric transmission factor K_{ij} is defined as the ratio of global irradiation at ground level on a horizontal surface (G_{ij}) to the insolation outside the atmosphere (G^{o}_{ij}):

$$K_{ij} = G_{ij} / G^{o}_{ij} \tag{3.17}$$

This quantity varies typically from 0.2 to 0.8. If one interprets the cloud cover index (n^{t}_{ij}) as the fractional coverage of a pixel (i,j) by clouds, the global irradiance at ground level and at time t is expressed as a linear combination:

$$G^{t}_{ij} = n^{t}_{ij} G^{1}_{ij} + (1 - n^{t}_{ij}) G^{2}_{ij} \tag{3.18}$$

where G^{1}_{ij} and G^{2}_{ij} are the global irradiance at ground level for cloud-covered and clear skies, respectively.

For each of these extreme conditions, one can define a transmission factor respectively K^{1} and K^{2}, which is supposed to be constant during the same hour. This realistic hypothesis leads to the following linear relation:

$$K^{t}_{ij} = n^{t}_{ij} (K^{2}_{ij} - K^{1}_{ij}) + K^{1}_{ij} \tag{3.19}$$

The Mêtérologie Nationale provided measurements of the hourly global radiation for 27 stations listed in Table 3.6. These measurements were reduced to transmission factors and compared to the closest 5x7 km pixel in the image with a location error better than one pixel (5 km).

3.4.5 Determination of hourly global irradiation

The previous step provides for each pixel (i,j) and time step t, the coefficients a_{ij} and b_{ij}, defining a linear regression between the transmission factor k^t_{ij} and the cloud cover index n^t_{ij} determined from the satellite.

Once the coefficients a_{ij} and b_{ij} are known at the pyranometer station locations, the spatial correlations between these coefficients are determined and a method of interpolation between the stations is applied in order to define the complete field of coefficients a_{ij} and b_{ij}. This determination is done during a training period. It is assumed that this relationship field is stable in time or slowly varying with a seasonal pattern. For a new image, it is thus possible to estimate the transmission factor:

$$\hat{K}^t_{ij} = a_{ij}\, n^t_{ij} + b_{ij} \qquad (3.20)$$

from which is computed the estimation of the hourly global radiation

$$G^t_{ij} = \hat{K}^t_{ij}\, G^{0t}_{ij} \qquad (3.21)$$

The test period of May 1979 was split into two parts for verification purposes:

- training period: 5 to 12 May 1979
- independent data set: 13 to 21 May 1979

Ground truth measurements of global hourly irradiation were used during the training period in order to determine the coefficients a_{ij} and b_{ij} for the 27 stations over France. The coefficients were then used for the prediction of global hourly irradiation during the estimated period for a pixel size 5 x 7 km.

For verification, the estimated and measured values were then compared for the period 13 to 21 May 1979 using a linear regression. The correlation coefficient between measured and estimated values was higher than 0.8 except for four stations (Carpentras, Pau, Montpellier, Marignane) and the average R.M.S. error is 42 J/cm^2 (0.12 kWh/m^2) (Table 3.6). Results are presented for the stations of Nice and Strasbourg in Figs. 3.15 and 3.16.

r = correlation coefficient

a,b = parameters to be estimated from measurements

= standard error of estimate (7 $cm^{-1}h^{-1}$)

1	1.20	-46.24	14.3
2	1.04	- 6.07	27.1
3	0.97	2.10	24.5
4	0.89	40.76	24.7
5	0.97	2.92	28.1
6	1.00	8.60	45.6
7	0.66	118.91	55.3
8	0.92	27.43	26.4
9	1.04	-24.71	50.1
10	0.97	- 4.78	69.2
11	0.97	10.80	36.8
12	1.10	-40.77	24.8
13	1.11	-69.91	72.9
14	0.94	30.93	26.3
15	1.20	-41.97	44.8
16	0.97	8.22	22.3
17	0.91	-10.62	36.3
18	0.97	-61.21	56.4
19	0.86	91.80	72.1
20	1.01	-29.27	55.1
21	1.18	-42.76	28.2
22	0.92	- 2.68	60.5
23	1.04	-11.15	41.4
24	0.89	7.27	57.2
25	0.91	- 0.29	36.0
26	1.18	-10.20	58.5
27	0.97	-16.97	28.3
Average	0.92	6.71	42.2

Table 3.6 Statistical results of comparison between satellite-derived and ground measured-global hourly irradiation using the 27 stations of Météorologie Nationale. Prediction was performed on the 13 to 21 May 1979 period using the 5 to 12 May 1979 period for training
(1 kWh/m^2 = 360 J/cm^2)

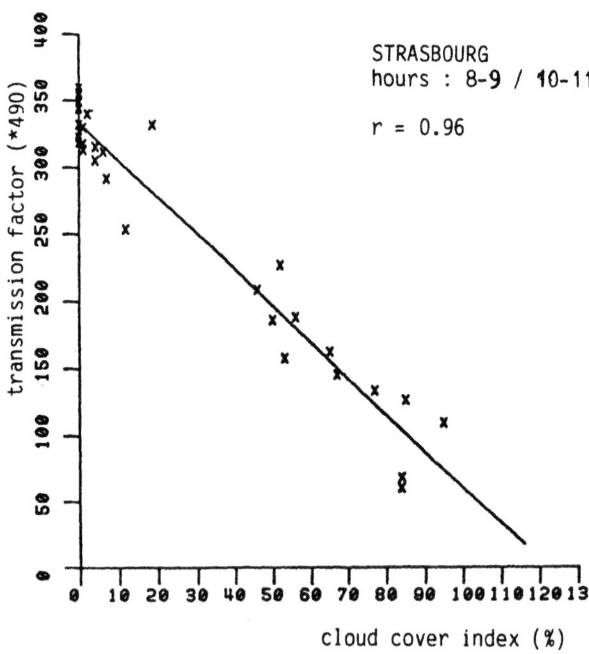

Fig. 3.15: Regression between satellite cloud index and transmission factor.

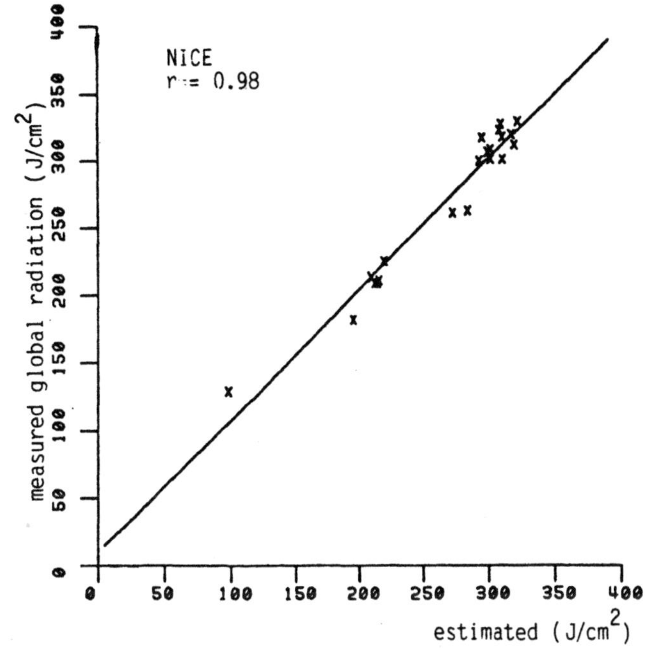

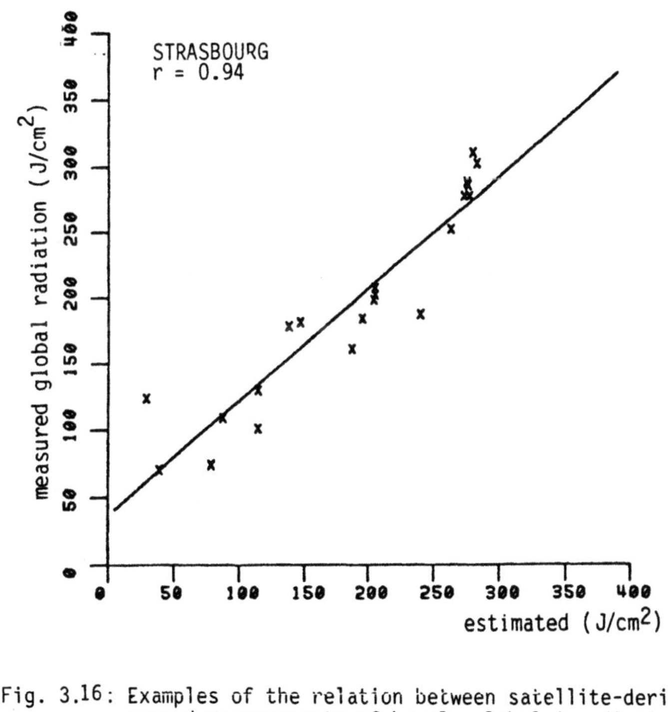

Fig. 3.16: Examples of the relation between satellite-derived and ground measurements of hourly global irradiation.

3.4.6 Conclusions

The hypothesis of linear correlation between the ground-measured transmission factor and the cloud cover index computed from METEOSAT is now verified and has been set up as the basis for the method used at Sophia-Antipolis.
If the ground albedo reference map has been constructed very carefully, then the use of a continuous cloud cover index provides an efficient and very simple method to estimate accurately the hourly global irradiation over Europe.

PART IV : COMPARISON OF THE THREE METHODS

4.1 Different approaches to satellite processing

In order to compute the solar radiation at ground level from satellite measurements, three types of approaches have been developed within this European project F (Action 4.3).

In the first case (see chapter 3.2) it is possible to build a physical model of the transmission properties of solar radiation through the atmosphere. It is used in an inversion-type approach whereby atmospheric properties are deduced from the Earth radiation measurements taken by the satellite scanning radiometer. Determination of parameters usually involves averaging the results over several pixels thus degrading the resulting resolution of the final product (example: 20 km).

It is also possible to rely on pure statistical relationships between satellite observations and solar radiation ground measurements. This method has been applied in Stuttgart (see chapter 3.3) based on multidimensional statistical techniques performed on texture analysis parameters extracted from the original satellite images.

Finally, it is also possible to use a combination of the two previous approaches in order to limit the number of regression parameters which have to be determined by statistical techniques. This is done by modelling part of the radiative transfer terms followed by statistical determination of the parameters which are difficult to model (see chapter 3.4).

This third type of method has also been used by the NOAA scientists (6) who have used multidimensional regression techniques in conjunction with a straight-forward expression of the atmospheric transmittance as a function of molecular scattering (Rayleigh) and water vapour absorption.

Data processing steps:

Whatever methodology was used by each of the research groups, data processing is organised in three broad steps.
First, some type of physical modelling is used at various levels of sophistication. This is summarized in Table 4.1. Then, the satellite data are preprocessed mostly for geometrical corrections and localisations and filtering of instrumental noise (Table 4.2). Finally, the data processing is performed with various degrees of statistical sophistication. (Table 4.3).

4.2 General background and rules of the intercomparison

The general scheme of an intercomparison was discussed and formulated at a meeting of all three groups held at Stuttgart on 15 and 16 October 1981. The final scenario and organisation were finalized later. The discussions and decisions were related to two topics: ground truth requirements and selection of a common test period.

4.2.1 Ground truth requirements

Until October 1981, the three groups had been acquiring their own ground truth on an individual basis. This was inefficient in more than one aspect. In particular, each group had to pay extra costs for copies, and also for conversion of the data into a format readable by the computer.

In order that all participants may use the same ground data set, it has been requested to provide a unified data set from the EEC countries. Certainly it would have been appreciated if the data from all European stations had been received in a short time. But the delay at the World Radiation Data Center in Lenin grad is about 18 months.

The possibility of obtaining data from a dense network was considered with interest as it lead to an assessment. The spatial resolution aspects of satellite measurements could then be made.

Table 4.1 Summary of major modelling features in the three models.

Groups from:	COLOGNE	SOPHIA-ANTIPOLIS	STUTTGART
Clear sky radiation	From calculations	From transmission coefficient	Estimates from measurements
Atmosphere	Calculations with models of aerosol and gas concentrations	Statistical estimates of linear transmission coefficients	-
Albedo	Maximum albedo for clouds; minimum albedo for clear sky	Separation of data between clear and cloudy conditions	Statistical separation of clouds
Clouds	Heights, shadows displacement and transmission	Transmission (statistical)	-
Geometry	Sun - satellite	idem	idem
Calibration of data	Solar channel: check of stability; infrared - use of data from ESOC	-	-

Table 4.2 Preprocessing techniques applied to raw satellite data.

Groups of	COLOGNE	SOPHIA-ANTIPOLIS	STUTTGART
Geometry	Polynomial + refinement (x,y, ,)	Translation landmark checking (x,y, ,)	Landmark selection resampling (x,y, ,)
Filtering	Average (6 x 8 pixels)	Filtering 5 x 5 pixels	Computation 64 x 64 pixels
Time interpolation	Linear between ground data	-	Linear between satellite data

Table 4.3 Comparison of data processing methods used in Action 4.3

Groups of	COLOGNE	SOPHIA-ANTIPOLIS	STUTTGART
determination of albedo	Clouds (absolute) ground	Clouds (relative) ground	-
Computation of image features	-	-	300 features
Determination of cloud cover index	Yes	Yes	-
Relation between cloud cover index and radiation	Physical, through radiation transfer modelling (compared with measurements)	Statistical, through physical regression (linear) Cloud cover/ radiation attenuation	Statistical, through multidimensional regression

Two participants proposed to carry out the data conversion from paper to magnetic tape as long as the amounts of data are not too large. The satellite data could be made available on an exchange basis; this was done later between the Cologne and Sophia-Antipolis groups. It was further agreed that hourly means would be sufficient. Any shorter time interval would lead to problems of availability and data handling.

4.2.2 Selection of a common test period and data

To avoid a dispersion of efforts (e.g. two participants independently selecting two consecutive 15-day periods for study in Spring 1979), it was agreed that the test should be carried out with a common test period. On the basis of availability of satellite and ground data, the month of April 1982 was selected. The ground data were provided in computer tape format by J.W. Grüter for a list of stations which are presented in Table 4.4. In this data set, only hourly averages of the global irradiance at ground as well as their daily sums were retained. After a first check of data consistency and quality, data from only 80 stations were used, which were mainly from Germany, France, Denmark, England and Ireland.

4.3 Goal of intercomparison between methods

The exercise of intercomparison has been arranged to ask for objective assessment of methods by using two periods, one for training of algorithms, the other for testing of performances. During this exercise, only capabilities of prediction at stations were tested, no test was performed for assessing the capabilities of methods in interpolating spatially between stations.

Country	Name	LAT	LONG	Country	Name	LAT	LONG
F	CLERMONT FERRAND	45.47	-3.10	B	OOSTENDE	51.12	-02.52
F	EMBRUN	44.34	-6.30	B	UCCLE	50.48	-04.21
F	HYERES-CEM-LEVANT	43.02	-6.28	B	SAINT-HUBERT	50.02	-05.24
F	MARIGNANE	43.27	-5.13	GB	LERWICK	60.08	+01.11
F	ODEILLO	42.29	-2.27	DK	HEDEN	57.23	-10.31
F	PAU	43.23	0.25	DK	AHOLM	57.16	-10.04
F	PERPIGNAN	42.44	-2.52	DK	JYNDEVAD	54.54	-09.07
D	LIST	55.01	-08.25	DK	HALD	56.34	-10.06
D	SCHLESWIG	54.32	-09.33	DK	FJALTRING	56.28	-08.08
D	BREMEN	53.03	-08.47	DK	GNIBEN	56.00	-11.17
D	NORDERNEY	53.46	-07.09	DK	SJAELSMARD	55.53	-12.25
D	HAMBURG-SASEL/FUHLSB.	53.39	-10.07	DK	BASTRUP	55.47	-09.39
D	OSNABRUECK	52.15	-08.03	DK	AVEDOERE	55.38	-12.26
D	BRAUNSCHWEIG	52.18	-10.27	DK	ARSLEV	55.19	-10.26
D	BOCHOLT	51.50	-06.32	DK	V. VEDSTED	55.18	-08.40
D	BRAUNLAGE	51.43	-10.37	DK	DUEODDE	55.00	-15.04
D	GELSENKIRCHEN/ESSEN	51.30	-07.05	DK	SARUPHAV	54.54	-10.03
D	KASSEL	51.19	-09.29	DK	ABED	54.50	-11.20
D	BONN-FRIESDORF	50.42	-07.08	IR	KILKENNY	52.40	+07.16
D	GIESSEN	50.34	-08.41	IR	VALENTIA	51.56	+10.15
D	COBURG	50.16	-10.57	GB	BRACKNELL	51.23	+00.47
D	GEISENHEIM	49.59	-07.58	F	SAINT QUENTIN	49.49	-03.12
D	WUERZBURG	49.46	-09.52	F	REIMS	49.18	-04.02
D	TRIER-PETRISBERG	49.45	-06.40	F	CAEN	49.11	+00.27
D	MANNHEIM	49.31	-8.33	F	PARIS/MONTSOURIS	48.49	-02.20
D	NUERNBERG	49.30	-11.05	F	TRAPPES	48.46	-02.01
D	SAARBRUECKEN	49.13	-07.07	F	NANCY/ESSEY	48.41	-06.13
D	WEISSENBURG	49.02	-10.58	F	STRASBOURG	48.33	-07.38
D	STUTTGART	48.50	-09.12	F	RENNES	48.04	+01.44
D	PASSAU	48.35	-13.29	F	AUXERRE	47.48	-03.33
D	WEIHENSTEPHAN	48.24	-11.44	F	TOURS/ST-SYMPHORIEN	47.27	-00.43
D	FREIBURG	48.00	-07.51	F	DIJON	47.16	-05.05
D	HOHENPEISSENBERG	47.48	-11.01	F	MACON	46.18	-04.48
GR	ATHENS (NAT.OBSERV.)	37.58	-23.43	F	LA ROCHELLE-PORT	46.09	+01.09
IR	BELMULLET	54.14	+10.00	F	LIMOGES/BELLEGARDE	45.52	-01.11
IR	DUBLIN AIRPORT	53.26	+06.14	F	AGEN	44.11	-00.36
IR	MALIN HEAD	55.22	+07.20	F	MILLAU	44.07	-03.01
IR	CLONES	54.11	+07.14	F	CARPENTRAS	44.05	-05.03
GB	ESKDALEMUIR	55.19	+03.12	F	NICE	43.39	-07.12
GB	ALDERGROVE	54.39	+06.13	F	MONTPELLIER	43.35	-03.58
GB	HEMSBY	52.41	-01.41	F	AJACCIO CAMPO D. ORO	41.55	-08.48
GB	ABERPORTH	52.08	+04.34	F	BISCAROSSE	44.26	1.15
GB	LONDON WEATHER C.	51.31	+00.07	F	BORDEAUX	44.50	0.42
GB	CAMBORNE	50.13	+05.19	F	CAPTIEUX	44.11	0.17
				F	CARCASSONNE	43.13	-2.19

Table 4.4 List of 80 pyranometer stations whose data were used in the intercomparison

Date	NOAA 6		NOAA 7		NOAA6
1.4.	6:15	7:56	13:11	14:53	17:40
2.4.		7:33	13:01	14:43	17:16
3.4.	7:09	8:57	12:57	14:38	
4.4.	6:46	8:34	12:45	14:26	18:11
5.4.	6:23	8:10	12:26	14:06	17:48
6.4.	6:00		12:14	13:54	17:25
7.4.	7:16	9:03	12:02	13:42	17:02
8.4.	6:53	8:40		13:30	18:18
9.4.	6:30	8:17		13:26	17:55
10.4.	6:07	7:54	13:14	14:56	17:32
11.4.		7:23		14:44	17:09
12.4.	7:00	8:47	12:51	14:32	18:25
13.4.	6:37	8:24		14:12	18:01
14.4.	(6:13)		12:20	14:00	(17:38)
15.4.	(7:30)		12:08	(13:48)	(17:15)
16.4.	7:06	8:54	13:36		(18:31)
17.4.	(5:43)		13:32		(18:06)
18.4.	6:20	8:07			17:45
19.4.		7:36	13:09	14:43	17:22
20.4.	7:13	9:00	12:49	14:31	16:59
21.4.	6:50	8:37	12:37	14:18	18:15
22.4.	6:24	8:13	12:26	14:06	17:49
23.4.	6:01	7:42	12:10	13:52	17:26
24.4.	7:27	9:06	12:09	13:50	17:02
25.4.	6:55	8:43		13:38	
26.4.		8:20	13:26	15:07	17:56
27.4.	6:07	7:49	13:03	14:55	17:32
28.4.	7:25		12:51	14:43	17:09
29.4.	7:01	8:42	12:39	14:31	18:26
30.4.	6:38	8:19	12:27	14:19	18:03

All times are in GMT

Table 4.5: List of satellite scenes processed in Stuttgart for the eintercomparison.

4.3.1 Background of method objectives

It was decided that the three methods would be tested with respect to their results on the global radiation at a horizontal surface at ground level in the form of hourly sums and daily sums of irradiation.It was also decided various capabilities. Thus while all groups calculated hourly sums, the groups from Cologne and Stuttgart determined also daily sums.

4.3.2 Intercomparison scheme

The month of April 1982 has been split into two periods:

(1) for training: 01 to 23 April 1982

(2) for test: 24 to 30 April 1982

The satellite data set was different for each of the groups. During this period the Meteosat radiometer showed serious malfunctions.

The METEOSAT data set has been provided by ESOC-Darmstadt in the form of:

- 73 VIS images
- 76 IR images

all stored on CCT

These observations were all obtained at three time periods of the day:

-08 h 53 GMT = Slot 18

-11 h 53 GMT = Slot 24

-14 h 53 GMT = Slot 30

The data set of NOAA satellites, on which the method of the Stuttgart group is based, was provided by the DFVLR-Oberpfaffenhofen or Dundee University, in the form of 122 images, spread around three time periods (Table 4.5):

-06 h 30 to 09 h 00 GMT

-12 h 00 to 14 h 30 GMT

-17 h 00 to 18 h 30 GMT

all on CCT

4.3.3 Form of presentation of final results

All results which were calculated from the satellite data taken during the test period (24 to 30 April 1982) were to be presented in two forms, displaying the statistics of the error of estimate.

$$\varepsilon = G_R - G_M \qquad (J/cm^2)$$

(= Predicted-ground measured by satellite)

Form 1: Histogram (Example: Hourly values)

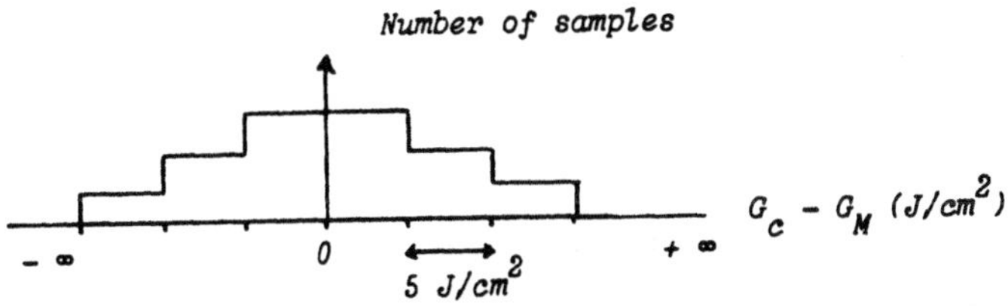

Form 2: Regression curve with statistics
(Example: Daily values)

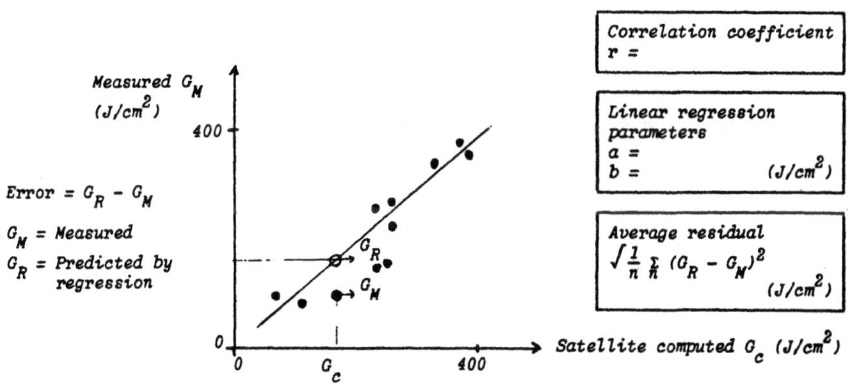

Fig. 4.1: Standardized presentation of results.

At a later meeting in Brussels (December 1982), the display diagram was considered as optional, comparisons being focused on numbers, namely:

- Correlation coefficient
- Average residual (J/cm^2).

4.3.4 Content of final report on intercomparison

Each group was required to produce a report, annexed to their results, containing information on the following problems:

- description of methods,
- deviation from ground reference data,
- hardware configuration,
- computational time:
  - satellite data processing to global irradiance image,
  - computation of global irradiation value on station basis.

4.4 Deviation between satellite-derived and ground-based radiation data

4.4.1 Reconstruction of hourly values

The error statistics are presented in the form of histograms of errors over the whole set of stations (80). A summary of results is given in Table 4.6:

Table 4.6 Standard error of estimate ($J\ cm^{-2}$)

	Time interval		
	08/09h	11/12h	14/15h
Cologne	20	28	21
Sophia-Antipolis	18	24	23
Stuttgart *	37.8	45.7	36.2

There is a remarkable consistency of results through the time periods. All the methods react to cloud fields, which form around noon.

* Obtained from digital images of the satellites NOAA 6 and 7.

4.4.2 Reconstruction of daily sums

The standard error of estimate for daily sums, as obtained by the two groups in Cologne and Stuttgart is given in Table 4.7:

Table 4.7 Standard error of estimate for daily sums (24-30 April 1983)

	in J/cm^2
Cologne	201
Stuttgart	244.6

4.5 Hardware configurations (as of 1982)

4.5.1 Cologne

Two types of computers have been used for this study:

- CDC Cyber 76 - University of Cologne
  Very powerful machine for radiative transfer calculations.

- Mulby 3/35 (KRANTZ) - 1974-model
  Mini-computer: 16 bits
  64 K bytes
  ADD in 15 μs
  MULT float in 30 μs

As far as the peripherals are concerned, the installation is autonomous and comprises:
- Tape drive: 9 T - 1600 BPI
- Disk drive: 48 M bytes

4.5.2 Sophia-Antipolis

Two types of computing devices are used:

- Array-Processor FPS-AP 120 B.
  High speed peripheral device used for determination of models (e.g.: albedo) and production of radiation maps (interpolation).
  Mini-computer: 16 bits
  64 K bytes
  ADD in 13 μs
  MULT float in 30 μs

- HP 1000

This group also is autonomous in peripherals:

- Tape drive: 9 T - 1600 BPI
- Disk drive: 20 M bytes

4.5.3 Stuttgart

This group also uses two types of computers:

- A host computer (16 bits)
- A special purpose hardware which has been designed by the group for high speed image processing.
  No information is available on the speed of this processor.

4.5.4 Image displays

For image manipulation, especially in the geometric correction and spectral training phases, each of the groups uses an image display processor:

- Cologne	256 x 256 pixels
	16 grey levels
- Sophia-Antipolis:	512 x 512 pixels
	256 grey levels
	+Graphic memory (512 x 512)
- Stuttgart :	512 x 512 pixels
	256 grey levels

4.5.5 Intercomparison

The hardware capabilities of each of the groups are very similar in the sense that:

- Image manipulation is interactively available,
- General purpose computation is done on 16 bits,
- Complex computations are done on specialized hardware:
- Radiation transfer (Cologne)
- Albedo model (Sophia-Antipolis)
- Image feature analysis (Stuttgart)

4.6 Intercomparison of computational time

Computational times as provided by the groups have to be looked upon as related to repetitive production after models and necessary training have been performed. In this intercomparison, however, one should bear in mind the different types of computers which were available to those groups at that time.

No time information was available for the group from Stuttgart. Time of processing is given per image. The times below refer to the equipment described above.

Cologne

- Pre-processing 1	12 min
- Quality control	10 min
- Pre-processing 2	20 min
- Boundary conditions	24 min
- Minimum radiance maps	16 min
- Maximum radiance function...........	5 min
- Radiation map production	30 to 120 min/map
- Output storage	30 min /map
TOTAL	147 to 227 min

for production of an hourly global radiation map on 2.5 x 2.5 km grid.

Sophia-Antipolis

- Pre-processing	3 min	
- Quality control.	3 min	
- Geometry and filtering	3 min	
- Cloud cover index	4 min	
- Radiation map production	4 min	(AP 120 B)
- Output	1 min	
TOTAL	18 min	

for production of hourly values on 5 km x 5 km grid.

It should be noted that both methods propose to map the final global radiation results on a refined grid (2.5 km or 5 km), thus interpolating between available ground stations.

4.7 Conclusions

The determination of global solar radiation at ground level using satellite data has been proven to be feasible by the three research groups which have participated in Project F of the European communities R + D programme. Whatever the methodology proposed whether based on physical modelling of the atmosphere or relying only on multidimensional regression techniques or even using both approaches together, the results are promising. The mean-square error of reconstruction of radiation at stations scattered over Europe has been demonstrated to be around 30 J cm^{-2} for hourly values and around 200 J cm^{-2} for daily values.

Following the research phase which has been supported by the European Communitiy Research and Development Programme, it is now a matter of turning satellite results into useful documents for solar meteorology. This has first been fulfilled by the participating groups in the form of publications of various satellite results obtained during the experiment.

These satellite-derived documents display in detail the microclimatic variations of the solar radiation over the land, particularly in mountainous areas. Apart from their potential use in solar engineering, these documents provide also vital new information over the oceans where, for the moment, almost no solar radiation map could be produced. These new data will certainly improve the global knowledge of the Earth´s radiation budget.

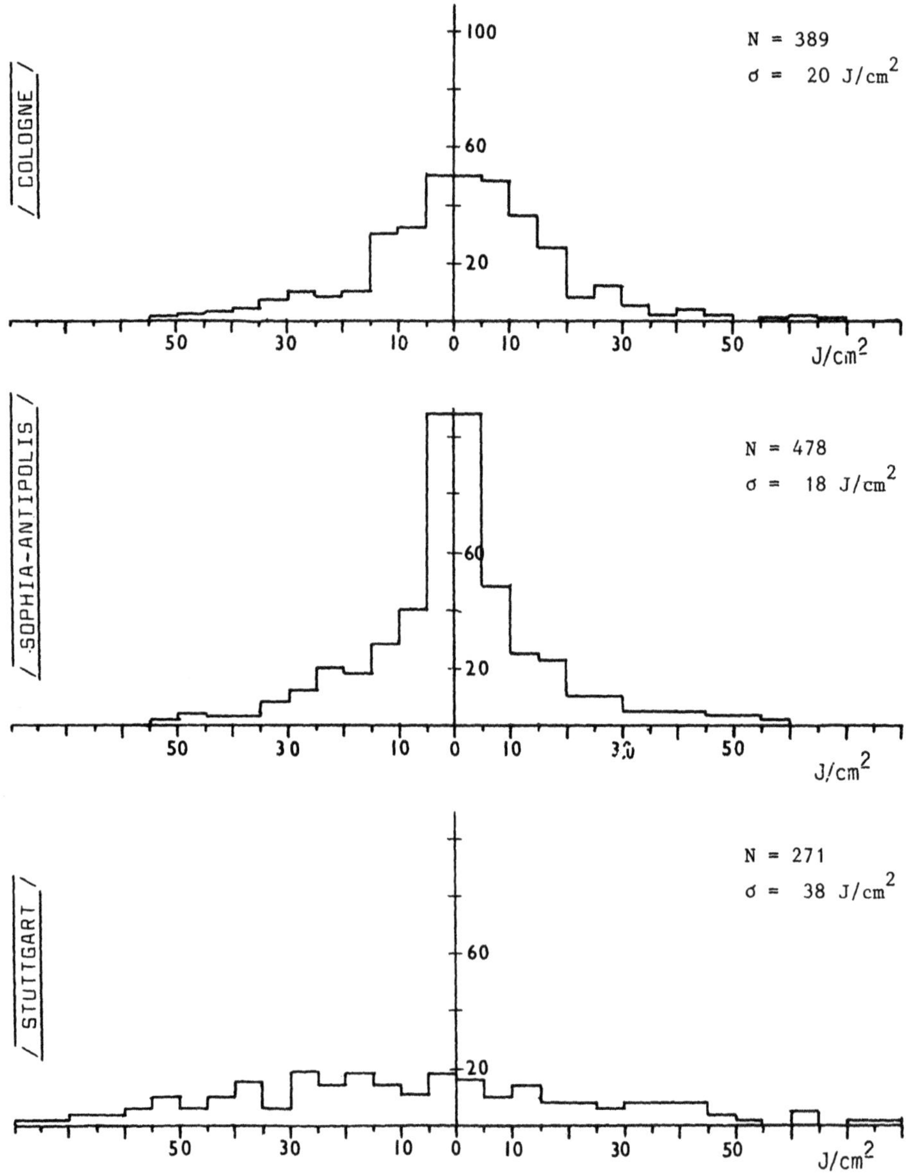

Fig. 4.2.: Histograms of hourly sums (about 8.53 GMT)

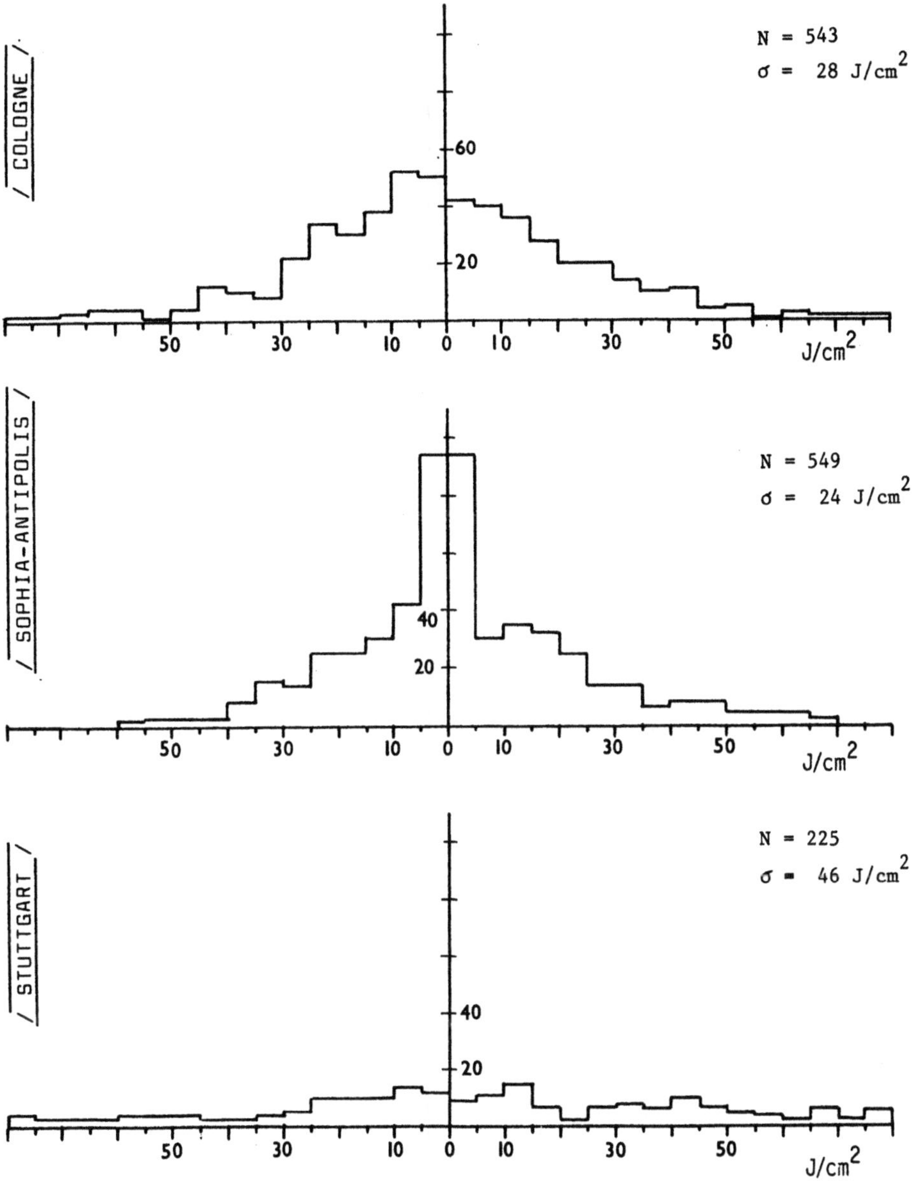

Fig. 4.3: Histograms of hourly sums (11 - 12 GMT)

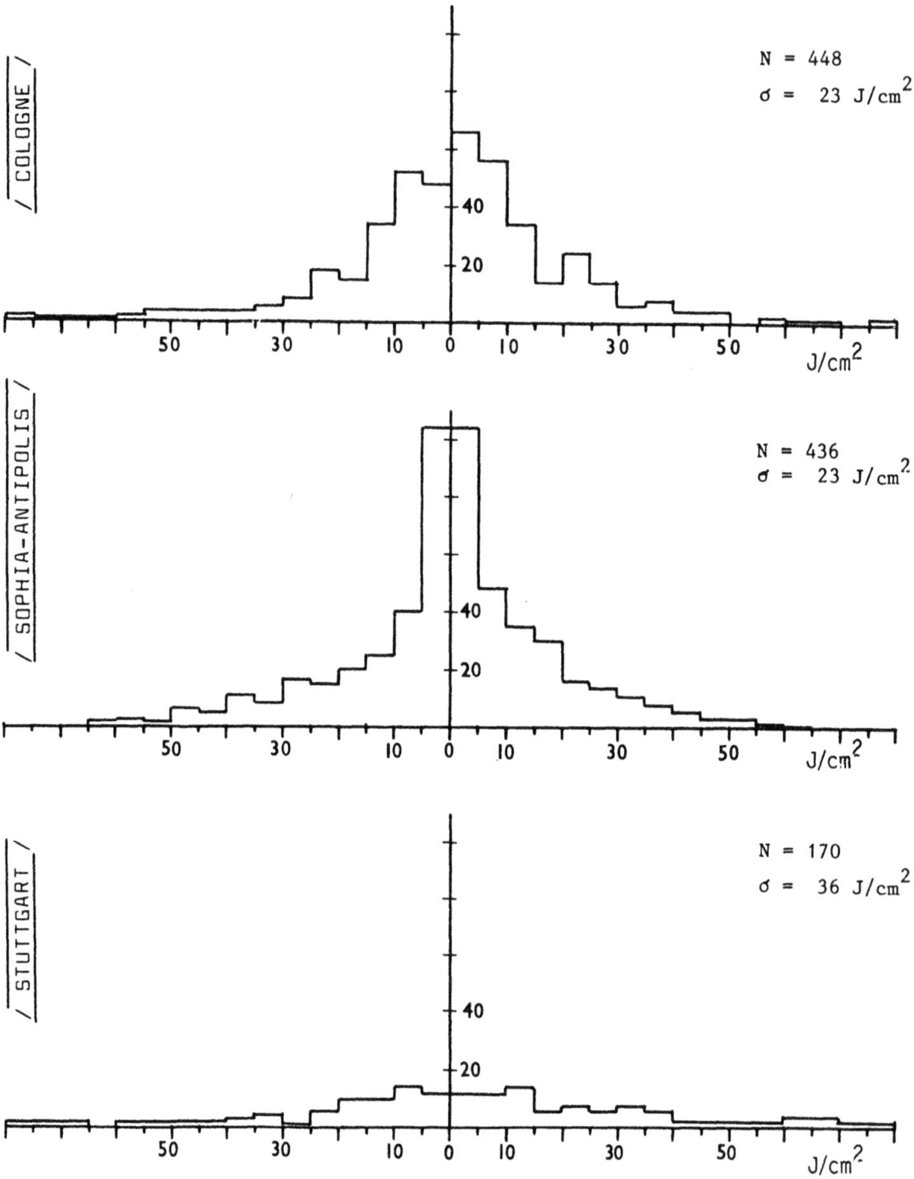

Fig. 4.4: Histograms of hourly sums (14 - 15 GMT)

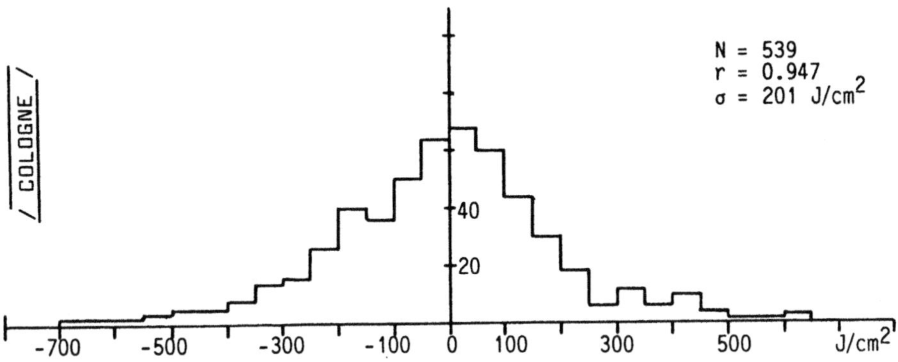

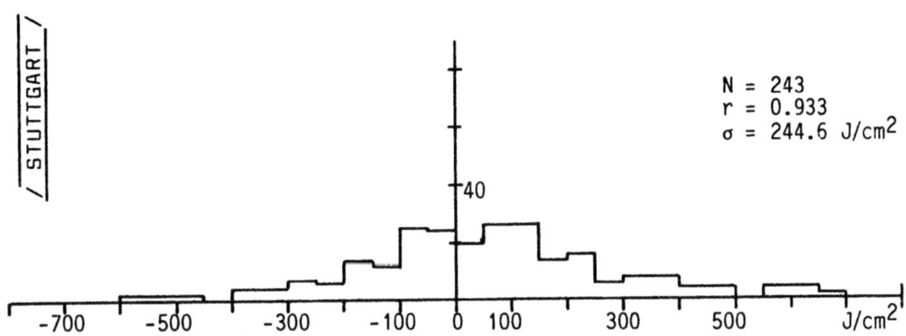

Fig. 4.5: Histograms of daily sums

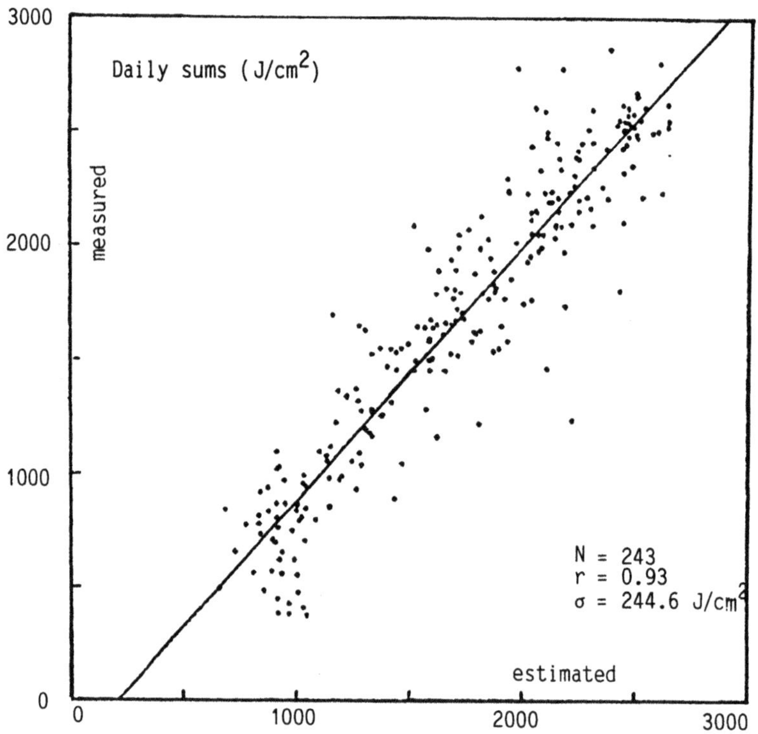

Fig. 4.6: Daily sums: correlation diagram of results obtained by Reinhardt et al. (Stuttgart)

PART V : GENERAL CONCLUSIONS

The determination of global irradiance or global irradiation at ground level using satellite data has been proven to be feasible by a number of authors and particularly by the three groups involved in Project F of the Commission of the European Communities R & D programme. So far only these three groups have undertaken an intercomparison of their methods. Whatever the methodology proposed, whether based on physical modelling of the atmosphere or relying only on multi-dimensional regression techniques or even using both approaches together, the results are promising. In this Action it has been demonstrated that hourly sums can be reconstructed from satellite data with an uncertainty of about 30 J cm^{-2} and daily sums with about 200 J cm^{-2} uncertainty. These values are smaller than 10 percent of averages; whereas the ground-based measurements themselves are already subject to uncertainties of about 5 percent.
It should be stressed that all these methods have in mind operational constraints: the volume of data to be processed is so huge that very detailed modelling of the physical properties of the atmosphere has been neglected in favour of simpler models in order to keep the computational time consumption of the method within tolerable limits.

These excellent results show clearly that satellite data can be used in a successfully to map the global radiation at ground level as well as the cloud coverage over very large areas. The ground resolution ranges from 5x5 to 50x50 km. The satellite-derived results display in great detail the microclimatic variations of the global radiation over the continent. These documents are of particular value for at least three applications. For countries where the solar radiation monitoring network is well developed, they offer a unique opportunity to interpolate between ground stations and hence to study microclimates. They also open up the possibility of mapping the solar radiation over countries where only a few measurement stations are available, as in Africa. Finally, they provide vital new information over the oceans where, for the moment, only a very few measurements are available and no global radiation map could be produced with reliable results. Indeed, these satellite-derived results should have a positive impact on the assessment of the Earth's radiation budget.

As for solar architecture, 15 to 50 % of the thermal energy demand can be drawn from solar energy, depending on seasonal variations. However, successful application depends on accurate dimensioning of storage and regulation devices. The design of engineering parameters for such devices calls for a precise knowledge of the amount of solar radiation incident onto a wall.

For concentration devices or photovoltaic energy-producing units, more details are needed in the radiation measurements with a time resolution better than hourly values. In this case one should also map statistics of clear sky occurrence and duration!

Solar climatology provides the necessary answers to these engineering problems through the modelling of variations of solar radiation in time and location at the surface of the Earth. It produces solar radiation atlases, which are used by solar engineers to tailor their computations to microclimatic variations due for example to variations of the optical transmission factor. These maps are sometimes printed in a form which permits the estimation of solar radiation over inclined surfaces with varying orientations.

Following the research phase, it is now necessary to apply these methods to larger data sets. Several other authors have also developed and used their schemes. Since a thorough knowledge of the available solar radiation at ground level is also required to study atmospheric and oceanic energetics, the National Aeronautics and Space Administration (NASA) is to hold a workshop in June 1985 where most of these methods will be the subject of thorough discussions.

However, some problems remain and must be solved before these methods can be applied world-wide. These problems are primarily concerned with the construction of maps of the ground albedo and of the upper limits of cloud albedos.

Over many seas of the world, covered seasonally or permanently by clouds, it is difficult to discriminate between clouds and cloud-free areas in the solar data of the NOAA - satellites or of METEOSAT. Therefore new approaches are required here, which may even make use of measurements in other spectral regions, such as channel 3 of the NOAA AVHRR (at 3.7μm).
But also more refinements are necessary to determine the background signal,

i.e. the reflected radiance at clear skies or the ground albedo, since over many areas, there are also clouds permanently present. Furthermore, the surface albedo of many continental areas undergoes seasonal changes. These uncertainties can be reduced only by the use of additional data; in some cases it might even be necessary tu use direct measurements on the ground.
Several authors have tried to determine the solar radiation over desert regions from Meteosat data. There, clouds play only a small role, and most of the attenuation is due to scattering and absorption by dust particles. In most cases, however, these can not be observed from space. Sandstorms cause usually observable signals, when their dust load passes from continental onto ocean surfaces.
The cases mentioned above might suffice to demonstrate the need for well controlled pyranometer networks on the ground. There is possibly no need to maintain them high density in all countries - which would be very expensive. But several small scale networks, covering at ahigh density an area of about 50 x 50 km would be most welcome for calibrating the results of global irradiance at ground level as derived from satellite imaging data. Such small networks would, moreover, contribute also to our knowledge of all energy transfer processes near the ground.

REFERENCES

Albuisson, M., H.Demarcq, J.M.Monget, N.Regas, L.Wald, 1984: Exploitation de la station de réception et de traitement d'images METEOSAT. Convention AFME N° 3.350.2797. Rapport intermédiaire CTAMN/83/R/05.

Amado, J., C.Delorme, P.Rabarento, 1981: Methologie d'utilisation des images du satellite METEOSAT pour la détermination du gisement solaire. Rapport GOMES - GDTA, N° 81877.

Ambrosetti, P., H.E.B.Anderson, L.Liedquist, C.Fröhlich, Ch.Wehrli, H.D.Talarek, 1984: Results of an outdoor and indoor pyranometer comparison Document No. III.A.3 of the IEA Solar Heating and Cooling Program.

Anonimus, 1978: Introduction to the METEOSAT - system, ESA METEOSAT Data Management Department, Issue 2.

Bourges, Juillet 1979: Courbes de fréquences cumuleés de l'irradiation solaire globale horaire recue par une surface plane. Rapport du Centre d'Energétique de l'Ecole Nationale Supérieure des Mines de Paris.

Cano, D., 1982: Etude de l'ennuagement par analyse de séquences d'images de satellite. Thèse de Docteur-Ingénieur, ENST.

Cole, A.E., A.Court and A.J.Kantor, 1965: Model atmospheres. Chapter 2 of the Handbook of Geophysics and Space Environments Air Force Cambridge Research Laboratories, Massachussetts.

Coulson, 1975: Solar and terrestrial radiation Acad. Press.

Dave, J.V., J.Canosa, 1974: A direct solution of radiative transfer equation: Application to atmospheric models with arbitrary vertical nonhomogeneities J. Atmos. Sci., 31, 1089-1101.

Dedieu, G., P.X.Deschamps, Y.H.Ken, 1983: Solar irradiance at the surface from METEOSAT - visible data. Machine processing of remotely sensed data symposium.

Dehne, K., 1980: Wichtige Spezifikationen von Pyranometern für Solarenergie--Belange Proc. 2nd Int. Solar Forum, Hamburg, 2, 163-178.

Dogniaux, et al.,1984: Solar meteorology (units and symbols), Recommendations by the solar energy R & D programme of the European Community Int. J. Solar Energy, Vol. 2, 249-255.

Ebke, W., 1982: Bestimmung der diffusen Solarstrahlung mit geeigneten Pyranometern, Thesis at the University of Cologne.

Gautier, C., G.Diak, S.Masse, 1980: A simple physical model to estimate incident solar radiation at the surface from GOES satellite data. J. Appl. Meteor., 19, 1005-1012.

Gautier, C., 1982: Mesoscale insolation variability derived from satellite data
J. Appl. Meteor., 21, 1, 51-58.

Hickey, J.R., B.M.Alton, J.F.Griffin, H.Jacobowitz, P.Pellegrino, E.A.Smith, 1982: Observation of the solar constant and its variations, emphasis on NIMBUS 7 results.
Proc. 3rd Scientific Conf. on Atm. Radiation, Davis (Cal), 164-166.

Howell, H.B., H.Jacobovitz, 1970: Matrix method applied to the multiple scattering of polarized light,
J. Atmos. Sci., 27, 1195-1206.

Gube, M., 1980: Outgoing longwave flux computation from METEOSAT data.
ESA - Journal, 4, 381-396.

Kasten, F., H.J.Golchert, R.Dogniaux and M.Lemoine, 1982: Global irradiation on ahorizontal plane. Definition of radiation climate zones.
Proc. of EC contractors' meeting, 18-19 Oct. 1982.
Solar energy R&D in the European Community, Series F.
Solar Radiation Data, Vol. 2, 50-62
D.Reidel Publishing Company, Dordrecht, Holland.

Kerschgens, M., U.Pilz, E.Raschke, 1978: A modified two stream approximation for computation of the solar radiation budget in a cloudy atmosphere.
Tellus, 30, 429-435.

Lacis, A., J.E.Hansen, 1974: A parametrization for the absorption of solar irradiation in the earth's atmosphere.
J. Atmos. Sci., 31, 118-133.

Liou, 1976: On the absorption, reflection and transmission of solar radiation in cloudy atmosphere
JAS, Vol.33, 798-805.

Möser, W., E.Raschke, 1981: Determination of global irradiation and of cloudiness from METEOSAT image data.
Colloque sur la Météorologie de l'Energie Solaire.
PIRDES - CNRS, Toulouse.

Möser, W., E.Raschke, 1983: Mapping of global radiation and of cloudiness from METEOSAT image data, Met. Rundschau 36, 33-41.

Möser, W., E.Raschke, 1984: Incident solar radiation over Europe estimated from METEOSAT data
J. Climate and Appl. Met., Vol. 23, No. 1.

Pastre, C., 1981: Développement d'une méthode de détermination du rayonnement solaire à partir des données de METEOSAT.
Preprint of la Météorologie VI, série N° 24.

Perrin de Brichambaut, C., G.Lamboley, 1974: Le rayonnement solaire au sol et ses mesures.
Cahier AFEDES, N° 1, Ed. complétés.

Plass, G.N. and G.W. Kattawas, 1968: Monte Carlo calculations of light scattering from clouds
Appl. Opt. 7, 415-419.

Plazy, J.L., P.Gregoire, J.McGregor, 1982: Calibration and characteristics of radiometers.
Proc. of EC contractors' meeting, 18 - 19 Oct. 1982,
Solar Energy R&D in the European Community, Series F:
Solar Radiation Data, Vol. 2, 2-8
D.Reidel Publishing Company, Dordrecht, Holland.

Raschke, E., H.J.Preuß, 1979: The determination of the solar radiation budget at the earth`s surface from satellite measurements.
Meteor. Rundschau, 32-18, 28 February 1979.

Raschke, E., 1972: Multiple scattering calculations of the solar radiation in an atmosphere - ocean system
Beitr. Phys. Atmos., 45, 1-9.

Schwalb, A.: The TIROS-N AG satellite series
NOAA Technical Memorandum. NESS 95.

Solar Energy R&D in the European Community, Series F: Solar Radiation Data,
Vol.1 D.Reidel Publishing Company,
Dordrecht, Holland.
Edited by W.Palz, 1981.

Tarpley, J.D., 1979: Estimating incident solar radiation at the surface from geostationary satellite data.
J. Appl. Meteor., 18, 1172-1183.

Tricaud, J.F., 1978: Atlas énergétique du rayonnement solaire pour la France.
CNRS - Pyc. Edition.

Tuzet, A., W.Möser, E.Raschke, 1984: Estimating global solar radiation at the surface from METEOSAT data in the Sahel region
J. Rech. Atmos., 18, 31-39.

WMO, 1971: Guide to meteorological instruments and observing practices.
WMO - N° 8, TP. 3, chap. 9, WMO Geneva.

GPSR Compliance

The European Union's (EU) General Product Safety Regulation (GPSR) is a set of rules that requires consumer products to be safe and our obligations to ensure this.

If you have any concerns about our products, you can contact us on ProductSafety@springernature.com

In case Publisher is established outside the EU, the EU authorized representative is:

Springer Nature Customer Service Center GmbH
Europaplatz 3
69115 Heidelberg, Germany

Batch number: 10584448

Printed by Printforce, the Netherlands